DEEPFAKES

ABOUT THE AUTHOR

Nina Schick is a broadcaster, author and advisor. She specializes in how technology and AI are reshaping politics. She has worked on Brexit, Emmanuel Macron's election campaign, on foreign interference in elections, including in the 2016 and 2020 U.S. elections, and on the evolution of mis- and disinformation.

Nina has advised a group of global leaders including Joe Biden and Anders Fogh Rasmussen (the former Secretary General of NATO), through her research on next-generation disinformation and AI-generated deepfakes. Half German and half Nepalese, she speaks seven languages and holds degrees from Cambridge University and University College London. She divides her time between London, Berlin and Kathmandu.

DEEPFAKES

The Coming
Infocalypse

NINA SCHICK

TWELVE

NEW YORK   BOSTON

Twelve
Hachette Book Group
1290 Avenue of the Americas, New York, NY 10104
twelvebooks.com
twitter.com/twelvebooks

Published in the UK as *Deepfakes and the Infocalypse* by Monoray
First U.S. Edition: August 2020

Twelve is an imprint of Grand Central Publishing.
The Twelve name and logo are trademarks of Hachette Book Group, Inc.

The publisher is not responsible for websites (or their content)
that are not owned by the publisher.

The Hachette Speakers Bureau provides a wide range of authors for speaking events. To find out more, go to www.hachettespeakersbureau.com or call (866) 376-6591.

Library of Congress Cataloging-in-Publication Data has been applied for.

ISBNs: 978-1-5387-5430-6 (hardcover), 978-1-5387-5431-3 (ebook)

Printed in the United States of America

LSC-C

10 9 8 7 6 5 4 3 2 1

CONTENTS

Introduction

"fucked–up dystopia"

There is a viral video of President Obama on YouTube, with almost 7.5 million views. The title lures you in: "You Won't Believe What Obama Says In This Video!" Obama looks straight into the camera. Seated in a deep mahogany chair, he appears to be in the Oval Office. He's aged—you can tell from his salt-and-pepper hair. But he looks confident, relaxed. Over his right shoulder, you catch a glimpse of the American flag. As usual, Obama is dressed impeccably: a crisp white shirt and a blue tie. On his left lapel, he's sporting a U.S.-flag pin. You click play. "We're entering an era in which our enemies can make it look like anyone is saying anything in any point in time," Obama begins. "Even if they would never say those things. So, for instance..." he continues, gesturing with his hands, "they could have me say things like President Trump is a total and complete dipshit!" His eyes seem to glimmer with a hint of a smile. Obama continues, "Now, you see, I would never say these things, at least not in a public address."

Obama never *did* say those things. The video was fake— a so-called "deepfake," created with the help of Artificial

Intelligence (AI). Welcome to the future, one in which AI is getting powerful enough to make people say things they never said and do things they never did. Anyone can be targeted, and everyone can deny everything. In our broken information ecosystem—characterized by misinformation and disinformation—AI and deepfakes are the latest evolving threat.

WHAT IS A DEEPFAKE?

A deepfake is a type of "synthetic media," meaning media (including images, audio and video) that is either manipulated or *wholly generated* by AI. Technology has consistently made the manipulation of media easier and more accessible (through tools like Photoshop and Instagram filters, for example). But recent advances in AI are going to take it further still, by giving machines the power to generate wholly synthetic media. This will have huge implications on how we produce content, communicate and interpret the world. This technology is still nascent, but in a few years' time anyone with a smartphone will be able to produce Hollywood-level special effects at next to no cost, with minimum skill or effort.

While this will have many positive applications—movies and computer games are going to become ever more spectacular—

it will also be used as a weapon. **When used maliciously as disinformation, or when used as misinformation, a piece of synthetic media is called a "deepfake."** This is my definition of the word. Because this field is still so new, there is no consensus on the taxonomy. However, because there are positive as well as negative use cases for synthetic media, I define a "deepfake" specifically as any synthetic media that is used for mis- and disinformation purposes.

The Obama YouTube video was produced by the Hollywood director Jordan Peele and Buzzfeed, and was intended to be educational—to serve as a warning for these potential negative use cases of synthetic media. As "Obama" goes on to say, "Moving forward we need to be more vigilant with what we trust on the Internet. It may sound basic, but how we move forward in the Age of Information is going to be the difference between whether we survive or if we become some fucked-up dystopia."[1]

Unfortunately, we are already in the "fucked-up dystopia." In the Age of Information, our information ecosystem has become polluted and dangerous. We are now facing a monumental and unprecedented crisis of mis-and disinformation. In order to analyse and discuss this problem, I needed to find a word to adequately describe this "fucked-up" information environment that we all now exist in. I settled on Infocalypse. For the purposes of this book, I define the Infocalypse as the increasingly dangerous and untrustworthy information ecosystem within which most humans now live

The word "Infocalypse" was coined by the U.S. technologist Aviv Ovadya in 2016, when he used it to warn about how bad information was overwhelming society, and asking whether there is a critical threshold at which society will no longer be able to cope. Ovadya used the term in relation to a collection of ideas, without settling on a single definition. However, as he correctly noted, the Infocalypse is not a static "thing" or one-off event, but rather an ever-evolving state of affairs, in which we all increasingly exist. It is my contention that the Infocalypse is evolving into an ever-more potent phenomenon with dangerous implications for everything from geopolitics to our individual lives.

It is difficult to pinpoint exactly when the Infocalypse came into being or to what extent it has taken hold. But it can certainly be linked to the exponential technological advances of the early part of this century. Before the turn of the millennium, our information environment evolved at a slower pace, one which gave society more time to adjust to technological advances. There were four centuries between the invention of the printing press and the development of photography, for example. But in the last three decades the

Internet, the smartphone and social media have transformed our information environment. By 2023, approximately two thirds of the world—5.3 billion people—will be plugged into this rapidly evolving information environment. The other third will soon join in. Video has emerged as the most powerful medium of communication in this ecosystem.

This rapid rate of change has made our information ecosystem ripe for exploitation. Increasingly, bad actors—ranging from nation states to lone "influencers"—are using this new set of circumstances to spread "disinformation"—or information that is meant to mislead—for their own nefarious purposes. Another side effect of this quickly evolving information environment is the spread of "misinformation." Unlike disinformation, which is meant to deceive, misinformation is simply bad information with no malicious intent behind it. Though neither mis- or disinformation is anything new, they have never existed at the scale that we currently face. They are also becoming more potent: in part, this is due to the emergence of mis-contextualised and/or edited video and photos, commonly known as "cheapfakes." Compounding this issue is the fact that we are still in the foothills of the AI revolution that is going to lead to a further evolution of our information ecosystem. As machines get better at generating synthetic media, the ways in which humans interact with one another and interpret information and the world will transform. Accompanying this AI revolution will be increasingly sophisticated mis- and disinformation in the form of deepfakes.

A notable feature of the Infocalypse is that it is becoming increasingly difficult to form a reasonable consensus on how to represent or perceive the world. All too often, it can feel as if one is forced to "choose a side." In the Infocalypse, even agreeing a framework of common "facts" within which reasoned debate can take place can be extremely challenging. As more and more people are becoming increasingly politicized in our polluted information ecosystem, well-intentioned efforts are directed into winning arguments over problems that become ever more intractable (race, gender, abortion, Brexit, Trump, Covid-19...), culminating in a doom-loop of partisanship. Neither side can persuade or convince the other in the Infocalypse—each attempt only risks entrenching further division. Ultimately, this growing divide in society will not be solved unless and until attention and energy can be redirected into addressing the structural problems of our broken information ecosystem. How did I come to be so interested in deepfakes and the Infocalypse? I saw it emerging as a consistent theme through my work in politics over the past decade.

THE INFOCALYPSE TAKES FORM

In 2014, I was working at an EU policy think tank in Westminster, analysing the EU's response to Russia's annexation of Crimea and recent invasion of Eastern Ukraine. I was working around the clock, contributing to one international news broadcast after another.[2] As the EU struggled to form its position, it became apparent that Moscow had a clear game plan. Russia simply denied that it had invaded Ukraine, claiming that Western politicians and commentators were waging an unjustified anti-Russia smear campaign.

Russia's version of events went something like this: Ukraine had descended into civil war, with pro-Russian "separatist rebels" fighting on one side and the Ukrainian state on the other. On one occasion, I had a memorable run-in with a pro-Kremlin commentator, an elderly gentleman who had once advised Vladimir Putin's predecessor, Boris Yeltsin. As conversations go, it was a disaster, although it might have made for great car-crash TV. We were not able to agree on simple facts, let alone have a proper debate. While I tried to explain the EU's response to Russian aggression, he denied that Russia was at war at all. With no shared reality, there was no basis for a sensible conversation.

For months, the crisis dominated my work. Events then took an even more surreal and tragic turn. The "separatist rebels" in Eastern Ukraine shot down a commercial airliner they had mistaken for a Ukrainian military flight. All 283 passengers and 15 crew onboard Malaysia Airlines Flight 17 (MH-17) were killed. While commenting on the Western response in broadcast studios across London, I was haunted by the footage accompanying the reports: the wreckage of that flight strewn across fields in Eastern Ukraine.

Subsequent investigations established that the Russian military was responsible for shooting down MH-17, even tracing how the missile launcher crossed the Russian border into Ukraine and back again.[3] To this day, Moscow still denies its involvement, even though we now know that this is a blatant lie. As the Intelligence and Security Committee in the British House of Commons concluded in 2017:

> Russia conducts information warfare on a massive scale [...] An early example of this was a hugely intensive, multi-channel propaganda effort to persuade the world that Russia bore no responsibility for the shooting down of MH-17 (an outright falsehood: we know beyond any reasonable doubt that the Russian military supplied and subsequently recovered the missile launcher).[4]

As these events unfolded, what was especially striking was how the Kremlin was capitalizing on the full range of emerging communication tools, especially social media, to disseminate its version of events. Russia Today (RT), the Russian state-sponsored international television network, was streaming its programming for free on YouTube. At the time, it was spreading pro-Kremlin narratives about MH-17 and the war in Ukraine. RT's editor-in-chief, Margarita Simonyan, gave an interview in 2014 in which she said that RT was "fighting" for Russia by "conducting the information war" against "the whole Western world."[5] She hit the jackpot by identifying an opportunity with YouTube. By 2017, YouTube was racking up over a billion hours of views per day on its platform. That is the equivalent of one person watching YouTube non-stop for 100,000 years.[6] And on that immensely powerful platform, RT is now the most watched news channel, with billions of views and programming in English, Spanish, French, German, Arabic and Russian. This is not television being made for the benefit of Russian ex-pats.

YouTube is not the only social media platform Moscow exploited. In 2013, the Kremlin set up the Internet Research Agency (IRA), as part of its intelligence services. Its mission was to use the social-media platforms to infiltrate public debate in foreign countries, and then "influence" them in a way that suited Moscow's objectives. The IRA targeted Ukraine first, but soon turned its attention to the West, infamously targeting the U.S. 2016 election. However, as I witnessed through my work, it was hitting Europe before that.

EXPLOITING THE MIGRATION CRISIS

Russia's disinformation initiatives in Europe are less well known, but I observed them unfold in the migration crisis of 2015–16. Russia's strategy was to harm Europe by instigating and exploiting a migration crisis. It did the former through physical warfare, starting with the acceleration of Russian military airstrikes to support the Assad regime in Syria in 2015. It was claimed that ISIS terrorists were the targets of the airstrikes, but the international community quickly established that what was happening was in fact an indiscriminate bombing of those civilians who were even moderately opposed to Assad.[7] NATO described it as a deliberate tactic to "weaponize migration" by unleashing a mass movement of people that would "overwhelm European structures and break European resolve."[8] Sure enough, after Russia's military operations in Syria, Europe's borders were soon overwhelmed by waves of mass migration (refugees, economic migrants and the odd terrorist). They largely arrived by sea. Many died. Fourteen thousand fluorescent-orange life jackets were later wrapped around the columns of a 19th-century concert hall in Berlin as part of an installation by the Chinese artist Ai Weiwei to commemorate the "drowned refugees."[9]

The EU member states were bitterly divided, provoked in large part by the German Chancellor Angela Merkel's open-door policy. More than a million migrants arrived in Germany in the summer of 2015. At the height of the flow, 10,000 people were entering Germany every day. They were unvetted. When Merkel realized the scale of the crisis, she did a U-turn and pushed for an EU-wide migrant quota to share the burden. The EU almost tore itself apart as some member states simply refused. Legal challenges were launched and borders were re-erected. New arrivals were directed to march on to Germany. Eventually, Germany was forced to seal its borders, too.

The implications of these monumental events have yet to be fully understood, but they will surely shape the EU (and Germany) for generations to come. The immediate political consequences have already transformed European politics. Islamic terrorists were among those travelling into Europe in the waves of migration, leading to several brutal terror attacks in major European cities, including Paris in November 2015, Brussels in March 2016 and Berlin in December 2016. Of 104 known Islamic terrorists who entered the EU between 2014 and 2018, 28 completed attacks that killed 170 and wounded 878. The majority of these terrorists applied for international protections such as asylum and were able to remain in European nations for an "average of 11 months before attacks or arrests for plots, demonstrating that asylum processes accommodated plot incubation."[10]

All the while, the Kremlin was pursuing information operations to pour further fuel on the already raging fire. Russia's information operatives infiltrated public discourse in European countries in an attempt to exacerbate building domestic tensions caused by the very migrant crisis it instigated. One such campaign was the story of "Lisa": a 13-year-old German girl who was allegedly gang-raped by refugees in Germany. The story was first reported on Russian national TV and then spread on social media. It quickly went viral, eventually leading to protests outside the German Chancellery in Berlin, with demonstrators accusing the government of a cover-up.[11] In reality, the story of "Lisa" was pure fabrication.[12]

The information space quickly degraded. It was filled with mis- and disinformation. The disinformation was so potent because it tapped into real and often legitimate fears, and misinformation was a further manifestation of those fears. Together, they created a febrile environment in which the public was bitterly divided. Many of the new populist European parties who benefited are believed to have links to the Kremlin, and their leaders have openly expressed sympathy towards Moscow, including by recognizing the annexation of Crimea and by calling for an end to EU sanctions on Russia. This includes Fidesz in Hungary; Lega Nord in Italy; Front National in France; the FPÖ in Austria (which was involved in a Russia-related scandal that led to the collapse of the Austrian government in 2019) and members of the far-right AfD in Germany.[13]

When I did some work on Emmanuel Macron's presidential bid in this feverish political atmosphere in 2017, for a while it looked like he might be beaten by Marine Le Pen. Two days before the election, Macron's campaign was then aggressively targeted by the same Russian hackers believed to be behind the hack of the Clinton campaign one year earlier.[14]

I also saw first-hand through my work on UK politics from 2013 onwards how these events were a key factor in the public's decision to vote for Brexit in 2016. Seizing on the tumultuous upheaval on the Continent, Brexit campaigners successfully made the argument that it was riskier to stay in the EU than to leave it. In my view, they had the decisive edge as they were able to appeal to public concerns on immigration. Seen through this lens, the EU migration crisis was a gift for Brexit campaigners. It made their messages even more powerful.[15]

A GATHERING STORM

The volatility of politics increased in tandem with the development the Infocalypse. This helped me see the potential of deepfakes as next-generation mis- and disinformation threats when I first encountered them in late 2017. Video and audio could now be generated or manipulated with AI. What was more, it soon became clear this technology would grow increasingly more accessible, and better, until anyone could use it. Everyone would have the power to show people in places they had never been, doing things they had never done, saying things they had never said. In the wrong hands, this technology posed a serious threat to our already corroding information ecosystem and, therefore, to how we understand and navigate the world.

In 2018, I started advising the former NATO Secretary General Anders Fogh Rasmussen as he was putting together a group of global leaders (including former U.S. Vice-President Joe Biden) to figure out how to build resilience to foreign election interference—especially with the upcoming U.S. 2020 election in mind. I knew deepfakes had be on the agenda: it was only a matter of time before they were used in the context of an election. As an advisor, I pushed these leaders as hard as I could to consider how AI might be exploited in the Infocalypse. I urged them to prepare for the inevitable attacks of the future.

Since then, the Infocalypse has become more potent. The upcoming U.S. presidential election in November 2020 will serve as a bellwether not only for the Western liberal democracies, but for the rest of the world. I am concerned that my warnings to Rasmussen et al. in 2018, about the threat of our corroding information ecosystem, will be proven prescient. Indeed, as I will address later, the polarization and distrust that characterize the Infocalypse are helping to perpetuate the social unrest that erupted in the United States just as this book went to press.

Since I started speaking and writing about mis- and disinformation and how deepfakes fit into this broader picture, the field has exploded. Although I first encountered deepfakes and our decaying information ecosystem through a political lens, it is clear that their impact will reach further than that. In writing this book, it is my modest aim to help you understand how dangerous and untrustworthy our information ecosystem has become, and how its harms extend far beyond politics—even into our private and intimate life. It is my hope that this understanding can help us come together to bolster our defences and start fighting back. As a society, we need to be better at building resilience to the Infocalypse. Understanding what is happening is the first step.

What follows is what you urgently need to know.

1

R/Deepfakes

Figure 1.1 Which faces are fake?

ook at Figure 1.1. Can you pick out which human faces are fake? If you chose the grainy black-and-white ones on the left, you are correct. But if you picked the ones on the right, you are also correct. All of these images are fake: synthetic media, generated by AI.

Until relatively recently, the manipulation of media—photos, video and audio—was the domain of specialists or those with immense resources, like a national government or a Hollywood studio. Technology is making human manipulation of media easier and more accessible to everyone. But, now, AI has granted humans a new tool by giving machines the power to generate wholly synthetic (or fake) media. This technology is still nascent, but we are in the early stages of an AI revolution which will completely transform representations of reality through media.

For the moment, the development of AI-generated synthetic media is outpacing society's understanding. We still tend to think of video and audio as authentic and incorruptible. As synthetic media become ubiquitous, however, we have to prepare for a world where seeing and hearing are no longer believing.

A HISTORY OF MANIPULATION

With the invention of photography in the 19th century, humans gained the ability to "capture reality" through a non-human medium for the first time. It quickly turned out that this medium could be manipulated. Photographic tampering has a long history. Take one early example from the 1860s. When Abraham Lincoln was assassinated, there was a dearth of "heroic-style" imagery of the President. To address this, one engraver decided to superimpose a photograph of Lincoln's head onto an engraving of the body of Southern politician John C Calhoun. For a century, no one noticed. Only recently was the print revealed to have been manipulated.[1]

Joseph Stalin is another famous name associated with the manipulation of photography. As terrible atrocities were committed in the name of Stalinism, his creed became synonymous with the rewriting of history, including of the visual record. An entire cottage industry dedicated to doctoring photographs developed under Stalin's dictatorship. Without modern photo-editing software, it required skilful mastery to be done well. Montages were composed by laboriously cutting up and stacking snippets of one photographic negative over another, for instance. Items were added through detailed hand-etching or removed by careful scratching of negatives. Stalin's pockmarked face was smoothed out in painstakingly slow and early versions of airbrushing. Stalin's Great Purges of the 1930s, in which he eliminated his political enemies, kept these artisans very busy. As the dictator's opponents were killed or sent to gulags on the orders of Stalin, they were simultaneously edited out of set-piece photographs. Take the example in Figure 1.2, on page 28. On the left, Stalin stands with a group of delegates at the Party Conference in April 1925. Six of the men later died by suicide, shooting or imprisonment, leading them to be "unpersoned," until only Stalin and three of his close friends remained in a version of the same photograph reproduced in 1939.

Figure 1.2 Top: Stalin stands with a group of delegates at the Fourteenth Party Conference in April 1925.Left to right: Mikhail Lashevich (suicide 1927); Mikhail Frunze (died 1925); Nikitich Smirnov (shot 1936); Alexei Rykov (shot 1938); Kliment Voroshilov (died 1969); Stalin; Nikolai Skyrpnik (suicide 1933); Andrei Bubnov (died in Gulag 1940); Sergo Ordzhonikidze (suicide 1937); Josef Unschlicht (shot 1938). David King Collection (TGA 20172/2/3/2/306)[2] Below: The same photograph reproduced in 1939 and retouched to leave only Frunze and Stalin's close friends Voroshilov and Ordzhonikidze.[3]

By 1990, as the Soviet Empire was breathing its last, photographic manipulation had started to become available to the masses through Photoshop, a commercial tool that democratized and improved on the manual work of the Soviet craftsmen and women under Stalin. Today, it is even easier to manipulate photographs. There is no need for an expensive computer or software. Anyone can download free and easy-to-use photo-editing apps onto their phone. We are wired to want to believe audiovisual material that "looks" or "sounds" right. Psychologists call this "processing fluency," referring to our unconscious cognitive bias in favour of information that our brain can process quickly. We do it a lot more quickly with visuals than with text. One study found that people were more likely to believe the claim "Macadamia nuts are in the same family as peaches" if the text is accompanied by an image of macadamia nuts, for example.[4]

Still, those of us who are more attuned to the fact that photographs can be manipulated can correct for processing fluency on the second or third view. This is not, however, the case for audio and video, which we still tend to see as incorruptible and authentic. In general, we tend to believe that video and audio footage captures what we would have seen with our own eyes or heard with our own ears, so they function as an extension of our own perception. This makes it even more worrying, then, that the means for AI-powered subversion of audio and video are developing at a time when these media are becoming the *most important* form of human communication—not only for the digitally savvy but for everyone. In the Information Age, we are not only mass consumers of audiovisual media, but also its *producers*. Billions of us now listen to and watch such media, but also document and share our lives through it using handheld devices. Synthetic media will take this even further, as Hollywood-level special effects will soon become accessible to everyone. It is estimated that by 2022, 82 per cent of global Internet traffic will come from video streaming and downloads. By 2023, over 70 per cent of the global population will have mobile Internet connectivity. Through their devices, 5.6 billion people will not only become consumers of online video, but producers too: they will not only listen and watch, but also record and share.

Synthetic media will take this even further, as Hollywood-level special effects will soon become accessible to everyone. This is an extraordinary development with unforeseen implications for our collective perception of reality. Subverting these media channels gives bad actors the tools to shape the perception of reality on scale that would have simply been unthinkable in the past. We are all vulnerable to this.

AT THE MOVIES

One place where we *are* conditioned to seeing manipulated audio and video is in movies, which we understand to be "make-believe." Visual manipulation in film is an art form stretching back to the birth of cinema, but by the 2000s it had become digital, with special-effects (SFX) experts working mostly with computer-generated images (CGI) for the best results. We have also witnessed a whole slew of commercial apps and software that have made video editing and special effects more accessible. However, the most powerful tools have remained in the hands of the well-resourced, such as film studios with multimillion-dollar budgets and teams of SFX experts.

AI is improving and democratizing for us all the tools that used to be reserved for Hollywood blockbusters. *The Irishman*, released in September 2019, is an on-screen adaption of the life of a mob hitman. Directed by Martin Scorsese, it features Robert De Niro, Al Pacino and Joe Pesci. The story spans seven decades, so the cast appear as young men in their twenties in the film. Scorsese used some of the $140 million budget to hire an SFX team to "de-age" his cast in post-production using CGI. He faced a series of technical challenges (including having to film with a three-lens camera rig) to make the de-ageing possible, which he later described as "just crazy."[5] For all his efforts, the end result doesn't quite work. "Their faces on screen perpetually look like they could be anywhere from 40 to 60 years old," wrote one reviewer of the clunky CGI. "It's confusing, and often distracting."[6]

Three months after the movie was released, a clip appeared on YouTube entitled "The Irishman De-Aging: Netflix Millions VS. Free Software!"[7] An anonymous YouTuber who goes by the name "iFake" used free AI software to have a crack at the de-ageing problem that had plagued Scorsese. Within seven days, iFake created an excellent result. The YouTuber's budget was zero and they were working alone. A side-by-side comparison of their work and the original film is available on YouTube. Their attempt is astonishingly good. Admittedly, iFake had a much easier challenge: the YouTuber was not working from scratch, using Scorsese's already heavily edited lens as their starting point. However, it is an early indicator of the power of synthetic media. Scorsese's production started in 2015. He had millions of dollars and a crack team of the best SFX artists. By December 2019, he was beaten by a lone YouTuber using free software. What had happened in the development of AI to make this possible?

AN AI REVOLUTION: DEEP LEARNING

AI was once the exclusive domain of science-fiction. Now it is a practical reality. It is going to change our lives beyond recognition, and synthetic media is only one manifestation of that. Computer scientists have been trying to figure out how to imbue computers with human levels of intelligence— or artificial intelligence (AI)—ever since their invention in the 1950s. By the 1980s, the AI community was split into two camps. One took a rules-based approach to developing AI, where computers would be programmed one rule at a time. The other camp hypothesized that the best way to develop AI would be to structure machines to learn *on their own* by mimicking the neural networks in the human brain to process data for decision making. This way, machines would learn through "experience" just like humans. This theory was not properly tested until the 2000s, when, for the first time, there was sufficient data and processing power to try it out. AI researchers found that it worked. By creating artificial neural networks loosely modelled on human brains, machines could indeed "learn," processing the data fed to them to perform tasks and make decisions autonomously. This so-called "autonomous machine learning" process, which relied on artificial neural networks, became known as "deep learning." In the last decade, the rapid acceleration in deep learning is catapulting the development of real-world AI forward, spawning numerous applications. Take facial recognition technology. Here, machine-learning systems are trained using a vast data set of human faces, until they autonomously learn to recognize them with perfect precision. Deep learning is also the foundation of the technology which pilots autonomous vehicles, where machine-learning systems are fed on vast data sets to train vehicles to detect and respond to stop signs, traffic lights and pedestrians.

AI research is characterized by its open-source and collaborative nature, meaning that cutting-edge research, tools and software are often shared freely on the Internet. In addition, because of the many positive applications of AI, the field is animated by private-investment academic research. This is generating a deep-learning revolution. And just as with all evolving technologies, this will be used for both good and bad. Such was the origin of the first deepfake. Someone figured out how to create a new type of video manipulation using emerging AI. It was a so-called "face swap." To make it, a type of machine-learning system known as an autoencoder was used to "teach" a computer to replace, or "swap," someone's likeness into an existing video. It all started on Reddit: ground zero for the synthetic-media revolution.

RULE #34

"AI-Assisted Fake Porn Is Here And We're All Fucked" announced the headline of a Motherboard article from 11 December 2017.[8] The author, Samantha Cole, was the first to break the news of deepfakes to the world. True to the "Rules of the Internet," porn is where the deepfake story begins. Originally written as a guide to protocols and conventions for those who identified with a group of online hackers known as Anonymous, the "Rules of the Internet" are now online folklore. There are many drafts, editions and disagreements over this list, which seeks to capture the essence of life online. In spite of this, there is almost universal acceptance of Rule #34: "There is porn of it. No exceptions." And so it goes that, when it comes to AI video creation, there is porn.

On 2 November 2017, an anonymous "redditor" called "Deepfakes"—a portmanteau of "fake" and "deep learning"—started a Reddit discussion forum: "r/deepfakes." He dedicated his forum to posting fake porn videos of Hollywood actresses, which he made using AI tools. He created these videos himself using open-source code that anyone with a working knowledge of deep-learning algorithms could put together. The circulation of fake (and leaked) celebrity porn is one of the Internet's favourite pastimes. But these creations demanded particular attention. As Samantha Cole later commented in a TedX Talk, they were different. They "moved, smiled, winced, and fucked."[9] This was a step up from a picture of Angelina Jolie's face photoshopped onto a naked porn star.

Early in 2020, I called Samantha to talk to her about her 2017 scoop. "It's my job to poke around obscure corners of the Internet," she told me. As a writer at Motherboard, the science and technology outlet for Vice Media, she is interested in the farthest corners of the web. Dystopian AI applications like deepfakes, YouTube feuds and questionable gadgets are exactly what interest her. She told me about how she was "poking around on Reddit" when she found the deepfakes forum, which was one of the top discussion threads on the website. "It was happening right under our nose," Sam explained. Like me, she was immediately struck by the potential wider implications of this technology. "If this one guy on Reddit could do this, what was stopping anyone from doing this to anyone?" she wondered. She messaged the creator, Deepfakes, who agreed to speak to her on the condition of anonymity. Deepfakes told Sam that he didn't consider himself a programmer or expert, but just someone who was "interested in machine learning." As Sam mused dryly on the phone to me, "We still don't know who he is, or what his real day job is." Deepfakes described how he had found a "clever way" to "face-swap" the faces of the celebrities onto the bodies of porn stars using open-source AI tools.

As with *The Irishman*, this type of media manipulation would have been the exclusive domain of SFX experts as recently as 2016. Putting together something like this would have required time, skill and money—significant barriers to entry. Deepfakes tore them down by using TensorFlow and Keras: open-source platforms for machine learning, the former developed by researchers at Google Brain, and the latter designed to allow fast deep-learning experimentation. Piggybacking off the work of these AI research superstars, Deepfakes launched his own Frankenstein's monster: an incest-themed video of the Hollywood actress Gal Gadot. As Sam described in her original article:

> There's a video of Gal Gadot having sex with her stepbrother on the internet. But it's not really Gadot's body, and it's barely her own face. It's an approximation, face-swapped to look like she's performing in an existing incest-themed porn video.

How did he do this? Using Google image search, stock photographs and YouTube, Deepfakes gathered a data set of Gal Gadot, which he then used to train an AI algorithm so that it would learn to "swap" Gadot's face—frame by frame—into an existing porn film. If you look closely at his creation for more than a few seconds, you can tell something is not quite right. There are glitches. Sometimes Gadot's mouth doesn't seem to sync in time to the words she is saying. When she is performing oral sex, a box appears around her head. But, if you weren't paying close attention, you could certainly be convinced that it was real.

Deepfakes openly shared his techniques on Reddit. Immediately, others started experimenting to make their own creations. No one was off limits. Two of the earliest targets for deepfake porn were superstar actresses who were introduced to the world when they were just children: Maisie Williams, best known as Arya Stark in *Game of Thrones*, and Emma Watson of *Harry Potter* fame. Cole's article exposing this activity on Reddit caused a furore. Within weeks, Reddit shut down the forum for containing "involuntary pornography." The person who had started it all, Deepfakes, disappeared, but not before he had openly shared the all-important "deepfake" code.

The "face-swap" techniques he pioneered were grasped by keen enthusiasts. New tools and free software dedicated to helping others make their own deepfakes soon started appearing on the Internet. Two of the most prominent of these are the software platforms DeepFaceLab and Face Swap. Run by mysterious programmers, they are dedicated free resources for anyone who wants to create their own synthetic media. Some degree of technical knowledge is required to use these platforms, but with patience and skill the results are excellent. These platforms have made anonymous YouTubers like "iFake," "Ctrl Shift Face" and "Shamook" famous. They use the free software for comedic purposes, swapping alternative actors into iconic Hollywood films. In one creation, Ctrl Shift Face replaces Macaulay Culkin in *Home Alone* with Sylvester Stallone, renaming it *Home Stallone.*[10] These videos are very funny and have racked up millions of views on YouTube. There are dozens of early deepfakes of Hollywood leading man Nicolas Cage. Thanks to his occasionally overly dramatic acting, Cage has become a beloved Internet meme. This has led to a movement to "face-swap Nicolas Cage into every movie ever made." One Youtuber, "DerpFakes," put together a "Nic Cage Mega Mix," featuring face-swap footage of Cage in movies from *The Sound of Music* to *Fight Club.*[11] These early uses are innocent and often really funny, but it is not typical. The most widespread use of faceswapping to date is to create non-consensual porn.

CATERING TO (MEN'S) WILDEST FANTASIES

From its birth on Reddit, deepfake porn has developed its own independent ecosystem. I learned more from speaking with DeepTrace, an Amsterdam-based company set up in 2018 to research deepfakes' evolving capabilities and threats. The firm is one of the first private companies set up expressly to help organizations and individuals protect themselves from the malicious use of AI-generated synthetic media. Where DeepTrace leads, many are bound to follow.

In a report released in late 2019, DeepTrace mapped the existing deepfake landscape. It found that deepfake content was on a quick upward trajectory, in part due to "the growing commodification of tools and services that lower the barrier for non-experts to create deepfakes."[12] By its count, there were almost 15,000 deepfake videos online by September 2019.

When I spoke to the founder of DeepTrace, Giorgio Patrini, he confirmed that the number of deepfakes (both porn and others) would go up significantly throughout 2020 and beyond. However, for now, they are still mostly in one category. DeepTrace found that 96 per cent of deepfake videos are of non-consensual porn. There is obviously a huge market for this. The first deepfake porn website was registered in February 2018. Now the top four deepfake websites have received over 134 million views. I visited one of these sites—easily found on Google—which proudly labels itself the "best source of DeepFake porn videos." All content can be accessed within a few clicks. All free. On the home page, the website welcomes viewers with the following:

> **There is a lot of porn out there. You can watch BDSM porn, Fetish porn, Foot fetish…Anal Sex, Blowjob Porn, Fisting…Teen porn you name it! Now imagine having those porn videos fakes using face-swap technology to insert a celebrity face onto it so it seems like she/he is the one performing all those sex acts![13]**

It goes on to boast how realistic these deepfakes are because they can "superbly capture" the emotions and expressions on celebrities' faces. Any fantasy can now be catered to like "a celeb porn video of Emma Watson giving a blowjob to a fan, or a NSFW celeb video of Scarlett Johansson having Anal Sex, or a celebrity fake of Maisie Williams having first time sex on camera." This technology can cater to even the wildest and "previously inaccessible fantasies"—for instance, the "fascination" of watching a popular figure such as "Ivanka Trump or Michelle Obama having BDSM sex on camera" while "Bill Clinton watches."

Deepfake porn is an undeniably gendered phenomenon. Of the hundreds of videos I scrolled through on the new deepfake porn sites, I did not see a single one of a male celebrity. No Brad Pitt, George Clooney or Johnny Depp. When I asked Sam Cole if she had come across any deepfake male porn during her time poking around in dark corners of the Internet, she shot straight back, "It's not something you see an awful lot of. No." But there are hundreds, literally hundreds, of female celebrities listed on these websites. Blondes, brunettes and redheads, women from dozens of countries, representing every ethnicity, and every celebrity you can think of. There is so much of it that it is even categorized: "Anal," "Threesome," "Masturbation" and so on. The website does not operate on any pretence that the porn is anything but non-consensual fake porn. A lot of the creations are "signed" by their creators.

Like the original Gadot video, many of the videos here can still be identified as fake, but they are getting better. In any case, the quality is beside the point. Any non-consensual fake porn, deepfake or not, high fidelity or not, is terrifying, embarrassing and demeaning to its victims. It is not clear how victims can protect themselves. This applies even to the wealthiest and best resourced of victims, such as Scarlett Johansson, the highest-paid actress in Hollywood. In a

2018 interview, she told *The Washington Post* that there is effectively nothing she or her team can do about it, because it is impossible to remove this type of content. "I have sadly been down this road many, many times," she said. "Every country has their own legalese regarding the right to your own image. So, while you may be able to take down sites in the U.S. that are using your face, the same rules might not apply in Germany." Frustrated, she called it a "useless pursuit" and warned that soon anyone could become a target, because the Internet is "a vast wormhole of darkness that eats itself."[14]

THE GANFATHER

Johansson's warning was prescient—what we have seen so far is just the tip of the iceberg. We are still at the very start of the synthetic-media revolution. The ways in which it can be generated are rapidly improving. The earliest deepfakes, face-swaps, are based on a class of deep-learning systems known as "autoencoders." However, a more versatile class of deep-learning systems is now taking over synthetic-media generation. It goes back to one man, the U.S. research scientist Ian J Goodfellow, and his invention in 2014 of a deep-learning system known as a GAN.

If you asked Siri to show you a picture of a computer

genius, it would give you Goodfellow. Bespectacled, goateed and with a mop of thick dark-brown hair falling over his forehead, the softly spoken Goodfellow looks exactly as you might imagine an "AI rock star" to look. One evening in 2014, Goodfellow sat down for a beer with friends at Les Trois Brasseurs, a popular drinking hole in Montreal where he was a PhD student in machine learning. They were discussing a deep-learning project they were working on and sharing how they were struggling to get their AI system to generate convincing-looking human faces. Early advances in deep learning meant that machines had become very good at categorizing data (the ability that underpins facial recognition, for example) but they still were not very good at generating it.

As Goodfellow listened and nursed his beer, he suddenly had an idea. What if, he wondered, you pitted *two* deep-learning networks against each in a game? One would try to generate new information and the other would try to detect it. As they played, the generator would learn to get better and better, in order to beat the detector. They would duel in a constant iterative process, until the generator could beat the detector. Essentially, he was applying the idea of "adversarial training" commonly used to train athletes to see whether it might work in deep learning. His friends scoffed, but when Goodfellow got home that night he got to work immediately. He programmed two deep-learning networks together into an adversarial game, to generate human faces. As the generator tried to beat the detector, it got better and better at the task. Goodfellow found that he had made an unbelievable breakthrough. Within hours, the system he had set up—the first generative adversarial network (GAN)—generated human faces that were better than anything AI had made before.

Look back at the synthetic-media faces this chapter began with. The black-and-white grainy ones on the left are the ones that Goodfellow generated that evening in 2014. Since his invention of GANs that night, the quality of their output has developed rapidly. The faces on the right were generated using a GAN just four years later in 2018. This should give you an idea of how radpidly the AI behind synthetic media is progressing. Thanks to GANs, AI can already generate close to perfect synthetic images. It is only a matter of time before the same is true for audio and video. Given the iterative and adversarial learning process used by a GAN, theoretically it will allow all AI-generated media to progress to the point where it literally becomes perfect. If you want to play around with how good AI is at generating wholly synthetic images, check out www.thispersondoesnotexist.com. Based on Goodfellow's research, it is a website that uses a GAN to generate new human faces. Every time you refresh the browser, you'll get a new fake AI person. You will be stunned at how even small details like wrinkles, pores and freckles are convincingly portrayed. But none of these people are real!

AN EXPANDING TECHNOLOGY

While GANs are the most exciting way to generate synthetic media at the moment, in the future researchers may develop even better ways to do it. There has been an explosion of research into synthetic-media generation. This incredible pace of development shows no sign of slowing, and is driven by the open-source nature of AI, but also by tremendous private investment. Given the many exciting and potentially lucrative real-world applications of synthetic media, there is a huge incentive to develop synthetic-media generation technologies until they become perfect.

Imagine what synthetic media will do for the film industry, for example. It will become easy to resurrect historical figures from the dead by training AI on their likeness. This will also be used for culture and art. The Salvador Dalí Museum in Florida used AI to bring the

Surrealist artist back to life in a synthetic-media installation called *Dalí Lives*, in which a video-generated Dali played "host" to visitors at the museum. Have a look at the video if you have the chance. It is an exciting and fascinating way to engage with the artist's body of work.[15] Synthetic media will also transform the gaming industry, making the visuals in games like *FIFA* so good that you could swear that it was the real athlete on your screen. In fashion AI is already being used to generate synthetic-media models. In August 2019, Zalando, a Berlin-based fashion and technology company, used a GAN to create synthetic images of AI-generated models in different clothing and poses.[16]

In the world of advertising, we saw the first synthetic-media advertisement in April 2020. It was broadcast to promote *The Last Dance*, the ten-part ESPN/Netflix documentary on Michael Jordan and the Chicago Bulls on their epic winning streak in 1990s. The ad begins with archival "SportsCenter" footage of the TV anchor Kenny Mayne reporting on the Bulls' sixth NBA championship in 1998. Suddenly, Mayne starts to go off-script: "This is the kind of stuff that ESPN will eventually make a documentary about," he says. "They'll call it something like 'The Last Dance.' They'll make it a ten-part series and release it in the year 2020. It's going to be lit. You don't even know what that means yet." Suddenly, a logo for State Farm, an insurance company, appears in the background. Mayne adds, "And this clip will be used to promote the documentary in a State Farm commercial."[17] It is an act of advertising genius. Viewers immediately fell in love with the clever concept. *The New York Times* wrote an article about the ad called "An ESPN Commercial Hints at Advertising's DeepFake Future."[18]

To understand more, I spoke to Victor Riparbelli, the CEO and Founder of Synthesia, a London-based start-up that is generating synthetic media for commercial use. According to Victor, video is by far the best way to deliver information, and synthetic media is a "glimpse into the future of how humans will create content." At Synthesia, he and his team are already generating synthetic media for their clients, who he says are mostly Fortune 500 companies. Victor explains that they are turning to AI-generated videos for corporate communications, because it gives them a lot more flexibility. For example, the technology enables Synthesia to produce a video in multiple languages at once. As the technology improves, Victor tells me, it will become ubiquitous. He believes that synthetic video may account for up to 90 per cent of all video content in as little as three to five years. This is great thing, he says. Today the creation of high-quality video content is "only for the people using high budgets and with connections to Hollywood," but when even a Youtuber can make high-quality video through synthetic media, it will allow "creativity to win."

Other data scientists whom I have spoken to think that we are about five to seven years away from AI being able to perfectly generate all forms of synthetic media for commercial purposes—and we are speaking about absolute perfection, given the exacting standards of these industries. Whatever the timeline, the point is that as the technology improves, it will obviously also be used for deepfakes. Victor concurs: "Whenever a new technology emerges, it will be used for bad."

At a time when video is becoming the most important medium of human communication, there is no doubt that deepfakes will be used as a weapon, taking video manipulation out of the realm of movies and into the real world. AI-generated synthetic media has three remarkable characteristics. First, there is its quality. AI is going to create audiovisual effects that are far better than anything CGI studios have done in the past. Second, there are its democratizing effects. As the technology improves, and it becomes more accessible via apps and software, more and more people will be able to access it. And third, there is its cost. As the technology improves, it is going to become cheaper or even free to produce this content. All of these trends are already under way.

At a time when audiovisual content is still highly trusted, and becoming ever more important, the malicious use of AI in the context of a rapidly corroding information environment has huge implications. Deepfakes did not start the Infocalypse, but they are the latest evolving threat. The consequences will play out on all levels of society—starting with geopolitics.

2

Russia

the master

Russian President Vladimir Putin is a real-life Bond villain. His staged photo opportunities, meant to showcase his masculinity, are, ironically, extremely camp. Here is the Russian President riding a horse bareback and topless; there he is pumping iron in the gym in a $10,000 outfit; or catching and kissing a "70 kg" pike. He might as well be stroking a white cat while he plots his global domination. No surprise, then, that with his reputation for being a "badass," President Putin has secured a place in the pantheon of Internet memes. However, these humorous associations with the image of Russia may lead us to underestimate the true threat the country represents in the Infocalypse.

Putin is one of the most dangerous men in the world. Over the last decade of his rule, Russia has started to enjoy an outsized geopolitical influence, partly because it uses the chaos of the Infocalypse to perpetrate ever-bolder attacks on the United States and the rest of the West. Russia was already a master of information warfare long before the Infocalypse came to define our information ecosystem. But by tracking three Russian disinformation campaigns against the United States from the Cold War to 2020, I hope to demonstrate how Russian attacks are becoming more dangerous in the Infocalypse. Moreover, Russia is also inspiring copycats, as other rogue and authoritarian nations seeking to exploit the conditions of the Infocalypse look to Moscow for inspiration.

COLD WAR ORIGINS

In 1984, Yuri Bezmenov, a high-ranking KGB defector, described Soviet disinformation techniques in a TV interview. Rather than "spying" in the traditional sense, he said that the KGB, the Soviet military intelligence agency, was almost entirely focused on using disinformation as a weapon of "psychological warfare," to distract and divide its Western enemy.

> [It] is a slow process which we call either ideological subversion or "active measures." [...] What it basically means is to change the perception of reality of every American to such an extent that despite the abundance of information, no-one is able to come to sensible conclusions in the interest of defending themselves, their families, their community and their country.[1]

His words have proven to be prophetic—now more than ever. But before we get to the Russian disinformation of the Infocalypse, let's look at one of the old Soviet campaigns, usually referred to as Operation Infektion.

OPERATION INFEKTION

In July 1983, an article entitled, "AIDS may invade India: Mystery disease caused by U.S. experiments" appeared in *The Patriot*, an obscure publication printed in New Delhi. The piece made a bombshell accusation: the deadly AIDS virus had been invented by the U.S. military as a biological weapon to kill black and gay men. The story cited a letter from an anonymous but "well-known American scientist and anthropologist" to support its explosive claims. Fort Detrick, a U.S. military base in Maryland, was at the centre of the allegations. During the 1940s Fort Detrick was home to the Pentagon's super-secretive biological weapons programme. A type of anthrax bomb was invented there during the Second World War. A million of them would have gone into production had it not been for the war ending in 1945. Another classified plan that never came to fruition at Fort Detrick involved spreading the yellow-fever virus by releasing infected mosquitoes from airplanes above enemy territory. Its labs had the capacity to produce half a million infected mosquitoes per month—a capability which was set to go up to 130 million per month. In this context, the AIDS accusation did not seem outlandish.

In reality, the U.S. biological warfare programme began to decline in the 1960s as President Nixon renounced the use of biological weapons. By the 1970s, Fort Detrick's remit had been scaled back radically to focus on defence against biological weapons, rather than their production. *The Patriot* reported otherwise, claiming that U.S. government scientists had scoured Africa and Latin America in secret missions to identify highly infectious pathogens, eventually leading to the creation of AIDS at Fort Detrick. The accusation of biological warfare was a trope that the Soviets used repeatedly against the Americans in the Cold War. (Make a note of this, as we will return to it in Chapter 6, when we examine the Covid-19 pandemic.)

For now, back to 1983, when the Soviets planted a seed of disinformation in a Soviet-sponsored Indian newspaper. It was before the birth of our modern information ecosystem, so this outrageous lie would have to be carefully cultivated to achieve virality. It took six years, but the Soviets managed to make it go global.

Here's how.

The *Patriot* article was published in 1983, making the claim that the Pentagon had created AIDS as a biological weapon. The myth went quiet for a few years, but the Soviets continued to accuse the U.S. of pursuing an offensive biological-warfare programme in violation of international laws. A 1985 Radio Moscow broadcast claimed that the CIA was spreading dengue fever in Cuba and helping South Africa develop a biological weapon to use against its black population. Suddenly, in 1985, the AIDS claim resurfaced in the influential Soviet weekly *Literaturnaya Gazeta*. In an article headlined "Panic in the West, or What is hidden behind the sensation about AIDS?" the journalist Valentin Zapevalov repeated the AIDS conspiracy claim, citing the "well-respected" Indian paper *The Patriot*. He obviously neglected to reveal that the story had been planted there by the Soviets.

The next year, a "scientific" report entitled "AIDS: Its Nature and Origin" by Professor Jacob Segal emerged, supporting the claim that AIDS was manmade. Upon closer examination, it turned out that Dr. Segal was a retired 76-year-old East German biophysicist. The co-author was his wife, Dr. Lilli Segal, also retired, an epidemiologist, and also East German.

As the AIDS epidemic worsened, the Soviets stepped up the campaign. Through 1986, a flurry of articles surfaced in the Soviet press repeating that the AIDS virus was created in a Pentagon laboratory. They widely cited the Segal report as scientific evidence, falsely reporting that Jacob Segal was a French (not East German) researcher. These reports were relayed globally by Soviet news agencies TASS and RIA Novosti, which together had over a hundred bureaus worldwide. The Soviets paid, tricked or otherwise incentivized local publications in the developing world to reprint the story.

By 1986, the story had started to go viral. It appeared in dozens of sympathetic as well as unsuspecting newspapers around the globe. This included the West, where it was printed on the front page of the British tabloid the *Sunday Express,* along with an interview with Jacob Segal. Segal's claims were soon debunked in Britain by conflicting reports in *The Times* and *The Sunday Telegraph*. Nevertheless, the operation was a great success: by the end of the decade the story had appeared in major newspapers in more than 80 countries. It was particularly virulent in Asia and Africa, where the lie received prominent play. The U.S. image was tarnished around the world.

Operation Infektion was particularly dastardly because it targeted the United States in a way that would divide it from within, too. By targeting the African-American community, the Soviets were aiming to exploit race relations and black Americans in the cruellest way. After all, it was easy for African-Americans to believe that their government wanted to murder them. They had already been experimented on by the government in the infamous Tuskegee experiment. Launched in 1932, when there was no effective treatment against syphilis, the study recruited 600 African-American men with the promise of free health care. They were told they were being monitored for "bad blood," a term used to indicate any number of health concerns. Instead, they were being studied to monitor the full progression of syphilis. None of the 400 men in the group who already had the disease were told. When penicillin became the recommended treatment for syphilis in 1947, they were still not told. When some passed the disease to their partners, and started having children with birth defects, they were still not told. When some went blind, became insane or died, still they were not told.

Operation Infektion took almost a decade to go global, but its legacy is still doing great harm. To this day, it is disproportionately believed in African-American communities, to the extent that it is still hindering HIV prevention among them. One survey of African-Americans found that 48 per cent believed that AIDS was an artificially made virus and 27 per cent believed it was made in a government laboratory.[2] As described by Yuri Bezmenov in 1984, it changed the "perception of reality" to such an extent that, "despite the abundance of information," Americans were not coming to "sensible conclusions in the interest of defending themselves."

The legacy of Operation Infektion even directly affected Barack Obama. It emerged that Obama's former pastor, the Reverend Jeremiah Wright, had claimed that the HIV virus was created by the U.S. government as a tool of genocide against black Americans and Obama was forced to publicly disown him. A seed planted by the Soviets in India in 1983 had come back to haunt the man who would become the first African-American President 25 years later. If the one lie at the heart of Operation Infektion lives on 40 years later, imagine the havoc that disinformation can wreak in the Infocalypse. Or what it could do when it is accompanied by deepfakes.

PROJECT LAKHTA

Russia's old Cold War techniques have become easier to deploy in the Infocalypse. During the last decade I have seen this unfold blow by blow through the invasion of Ukraine, the shooting down of Flight MH-17 and the EU migrant crisis. But even I was not prepared for the scale of what was coming next in 2016—one of the most brazen assaults so far, a coordinated, direct attack on U.S. democracy in the context of a deeply controversial American presidential election. This went far beyond the spreading of a single lie to hurt the United States, but, just as I had seen in Europe, Russia simply denied any involvement. The fact that this matter has become a partisan political issue in the United States, with one side paranoid about Russia and the other denying that Russia is a threat at all, shows that the Kremlin's strategy is working beautifully. I don't think they could have expected such "success" in their wildest dreams. And the Infocalypse made it so much easier.

The Kremlin-led attack is an incontrovertible truth. It should not be a partisan question. It has been established by all U.S. intelligence agencies. In a joint statement in 2017, the Central Intelligence Agency, the Federal Bureau of Investigations, the National Security Agency and the Office of the Director of National Intelligence said:

> **Russian efforts to influence the 2016 presidential election represent the most recent expression of Moscow's longstanding desire to undermine the U.S.-led liberal democratic order, but these**

activities demonstrated a significant escalation in directness, level of activity, and scope of effort compared to previous operations.[3]

They add that the attack was sanctioned by Vladimir Putin personally, and that he had a preference for Donald Trump over Hillary Clinton. "We have high confidence in these judgments."[4]

This type of attack would not have not been possible without the technology that has come to define our modern information ecosystem, and we now know that it was comprised of three strands, which included:

1 The hacking of voting systems;
2 The hacking of the Democratic National Convention (DNC) and Hillary Clinton campaign;
3 Using the Internet Research Agency (IRA) to run disinformation operations on social media to distract and divide American citizens.

To demonstrate how Russia's disinformation strategies have become more potent in the Infocalypse, I will focus here on the third strand: what the IRA did. As mentioned in the Introduction, the IRA was set up in 2013, first cutting its teeth during the invasion of Ukraine in 2014. In the United States, its operations would become known as Project Lakhta.

Just like Operation Infektion, the aim of Project Lakhta was to weaken the United States by using information operations. In 1983, Operation Infektion centred around one lie, but Project Lakhta was more ambitious in scope. The IRA was charged with infiltrating U.S. public discourse by posing as authentic Americans on social media, and then corrupting it by sowing as much discord, polarization, division and disinformation as possible. They would do this via social-media platforms: Facebook, Twitter and Instagram. They had a long-term game plan. As Special Counsel Robert Mueller found in his subsequent investigation, the IRA started its activities in the United States in 2013—three years ahead of the election.[5] In mid-2014, IRA agents even travelled to the United States in an "intelligence gathering mission [...] to obtain information and photographs for use in their social media posts."[6]

From their base in St. Petersburg, IRA agents next built fake pages, communities and personas, made to look as though they were authentically American. Then, this small group of operatives launched a multi-year plan to systematically corrupt American public discourse. This was about not only exploiting existing divisions as in Operation Infektion, but actively trying to create new ones. And the

IRA did it by playing with identity politics. It involved a two-pronged strategy. First, the IRA would build up an identity by constructing fake online communities around those identities. Then they would imbue these communities with positive messages so that they would feel pride in their distinct identity. This would strengthen feelings of togetherness and group identity. Then, once these identities had been built up into "tribes," they would inject the same communities with negative messages about other tribes to make them feel alienated from them.

The IRA intervened on all sides of the political divide. On the left, they toyed with the LGBTQ community and African Americans (again). On the right, they experimented with Texan secessionists and gun owners. Although the conventional narrative says that only "dumb Trump voters" were fooled, in fact everyone was. The Mueller Report later described the strategy in more detail:

> **IRA Facebook groups active during the 2016 campaign covered a range of political issues and included purported conservative groups (with names such as "Being Patriotic," "Stop All Immigrants," "Secured Borders," and "Tea Party News"), purported Black social justice groups**

("Black Matters," "Blacktivist," and "Don't Shoot Us"), LGBTQ groups ("LGBT United"), and religious groups ("United Muslims of America").[7]

By fostering tribalism, the IRA was hitting the United States in a weak spot: using its free and open democracy to exploit divisions in society. As noted by the Pentagon, Russia's tactics are "most effective when the target is deeply polarized or lacks the capacity to resist and respond effectively to Russian aggression."[8] By dividing and destroying the United States from within, the Russians hoped to immobilize it, and thus get away with their bare-faced assault.

It was fairly easy for the IRA to lure American citizens into its social-media networks. Using paid-for social-media advertising, which can be targeted due to the personal data points the big platforms collate on their users (from demographics to political affiliation), they could ensure its content was promoted to those whom they wanted to ensnare with almost pinpoint precision. They also experimented with creating content that would be so popular that it would go viral organically.

The Democratic House Intelligence Committee publicly released a cache of IRA social-media advertisements on Facebook between 2015 and 2017. They were prolific. On

Facebook alone, the IRA purchased 3,393 ads, with more than 11.4 million Americans exposed to them. They built 470 Facebook pages, and created 80,000 pieces of organic content, to which over 126 million Americans were exposed.[9]

Let's look at one IRA Facebook group, "Black Matters." As the name suggests, the group was aimed at African Americans. It gives the impression of being affiliated with Black Lives Matter (BLM), the American campaign group founded in 2013 to protest against the violence and racism to which black Americans are subjected, especially the police killings of African Americans. A paid ad sponsoring the IRA's "Black Matters" page goes straight for the jugular (Figure 2.1). It calls on people to "like" the page, saying "Join us because we care. Black Matters!" The ad features the faces of three young African Americans—Michael Brown, Tamir Rice and Freddie Gray—along with a banner reading in capitals "NEVER FORGET." Brown, Rice and Gray were all killed in police shootings or in police custody. Rice was only 12. His crime: he had been playing with a toy gun. He was shot on sight. By playing with such powerful tribal sentiments, the IRA was able to grow its influence operations. As in Operation Infektion, the IRA's disproportionate focus was on the black community because race is such a politically and socially sensitive issue in America. It also did it with the hope of supressing black

Figure 2.1 "Never forget"—paid ad from the IRA's "Black Matters" Facebook group

Democratic voter turnout for Hillary Clinton. As the election drew closer, the IRA bombarded these groups with selected narratives such as that Hillary Clinton did not care about black people, and that the race between Trump and Clinton was one that black people should be sitting out. Not all content from Project Lakhta was pro-Donald Trump. On the left, there was also pro-Bernie Sanders content. But it was all, universally, anti-Hillary Clinton.

The IRA targeted conservative voters, too. Black Lives Matter has become a highly visible, but not uncontroversial, group in America. It has spurred counter-movements, such as Blue Lives Matter, to support police, for example. In a sponsored ad for another one of its pages, "Being Patriotic," which is targeted at white, conservative voters, the IRA urges people to "like" its page by pushing Blue Lives Matter content. The ad shows an image of a police funeral, which is overlaid by text blaming a Black Lives Matter activist for "another gruesome attack on police." The ad also attacks Hillary Clinton, warning that she is the "main hardliner against cops," and that Donald Trump is the "one and only who can defend the police from terrorists."[10]

The IRA found that it could even get Americans to attend real-life political rallies, organized and promoted by the IRA from St. Petersburg via social media. The Mueller Report describes how:

[The IRA agents] sent a large number of direct messages to followers of its social media accounts asking them to attend the event. From those who responded with interest in attending, the IRA then sought a U.S. person to serve as the event's coordinator. In most cases, the IRA account operator would tell the U.S. person that they personally could not attend the event due to some pre-existing conflict or because they were somewhere else in the United States. The IRA then further promoted the event by contacting U.S. media about the event and directing them to speak with the coordinator. After the event, the IRA posted videos and photographs of the event to the IRA's social media accounts.[11]

As the election crept closer, the effort to get real Americans at such political events stepped up. The Trump campaign even ended up unwittingly promoting a fake IRA event on Trump's Facebook page.[12] But the IRA also ran anti-Trump rallies after his election! One of them, organized by the IRA's Black Matters U.S. page, managed to convene 5,000–10,000 protestors in Union Square in Manhattan. The angry rally then marched to Trump Tower to protest against his victory, four days after he won in 2016.[13] "Join us in the streets! Stop Trump and his bigoted agenda!" said the Facebook event page for the rally. "Divided is the reason we just fell. We must unite despite our differences to stop HATE from ruling the land."[14] The tragic irony, of course, being that division was exactly what the Russians were hoping to create.

Project Lakhta evolved constantly over the years as IRA agents experimented with what went viral. Renée DiResta, a leading disinformation expert from the Stanford Internet Observatory, describes one such example. She says that the

IRA's "Army of Jesus" page was originally created as a page dedicated to Kermit the Frog—a character from the children's programme *Sesame Street*—and then went on to become a meme page dedicated to *The Simpsons* cartoon. As they threw things at the wall to see what would stick, the IRA eventually settled on Jesus, finding that when they pushed ads urging Americans to "Like for Jesus" or "Share for Jesus," the content they were producing on these pages would go organically viral. Renée says the "Army for Jesus" page became "a powerful anti-Hillary meme factory in the run-up to the election."[15] One of the ads pushed from the page on the day of the election features Jesus arm-wrestling with Satan, with the following sprawling text: "SATAN: IF I WIN, CLINTON WINS! / JESUS: NOT IF I CAN HELP IT! / PRESS 'LIKE' TO HELP JESUS WIN!" (Figure 2.2).[16] The meme is accompanied by the following text:

> Today Americans are able to elect a president with godly moral principles. Hillary is a Satan, and her crimes and lies had [*sic*] proved just how evil she is. And even though Donald Trump isn't a saint by any means, he's at least an honest man and he cares deeply for this country. My vote goes for him![17]

Figure 2.2 Ad from the IRA's "Army of Jesus" Facebook group

The IRA found that memes were effective as they went viral easily. We tend to think of memes today as funny images online, but the term was coined by Richard Dawkins in his book *The Selfish Gene* (1976), where he described how culture is passed on through generations. By his definition, memes are "units of culture" spread through the propagation of ideas. Online, memes are a very effective mode of communication. Because they are seen as harmless, they often become unmoored from the context of their creation, allowing them to take on a life of their own, meaning that no one is responsible for any hateful or transgressive ideas contained within them. But as the IRA's influence operations have established, memes can be used as a weapon of information warfare. As Joan Donovan, a Harvard professor studying media manipulation and disinformation, wrote in the *MIT Technology Review* in 2019, memes "make hoaxes and psychological operations easy to pull off on an international scale." Meme wars have become a "consistent feature of our politics," as they are now being used by "governments, political candidates, and activists across the globe."[18]

The IRA also used its social media networks to prepare the ground for other parts of its attacks on the United States. For example, after Russia hacked the DNC and Clinton campaign and was preparing to leak emails onto the "hacktivist" WikiLeaks website, the IRA would pump its social-media networks full of positive messages about WikiLeaks and its founder, Julian Assange, such as that he was a hero and truth-teller. When the email dump came out days later, Americans would be more receptive to its contents.

When the truth about Russian interference in the U.S. election started to become public knowledge in 2016, the Kremlin simply denied it. Meanwhile, the IRA continued its social-media operations. It even started poking fun at the idea of Russian interference. One viral meme featuring President Barack Obama made an attempt to laugh off as ridiculous the claim of Russian interference (Figure 2.3). He is flanked by bearded Arab-looking men. Playing on conspiracy theories that Obama was in cahoots with the Muslim Brotherhood, the transnational Sunni Islamist organization originally established in Egypt, it reads (in capitals), "IF ONLY MEDIA HAD BEEN AS BOTHERED BY OBAMA'S TIES TO THE MUSLIM BROTHERHOOD AS THEY ARE BY TRUMP'S FAKE TIES TO RUSSIA." It then urges viewers to "SHARE IF YOU AGREE." The meme is clever because it plays into the widespread distrust of the media that has come to dominate the American political arena. It implies that the mainstream media is naive and biased for believing the "fake ties" between Trump and Russia, but dismissing the

Figure 2.3 IRA viral meme featuring President Barack Obama

purported links between Obama and the Brotherhood.

The IRA's operations continued well into 2017. It was only months after the election, when Russian interference had become one of the biggest political issues in American politics, that the IRA networks were weeded out and shut down. Disinformation researchers at the University of Oxford, who prepared one of the first reports into what the IRA was doing, found that

> surprisingly, these campaigns did not stop once Russia's IRA was caught interfering in the 2016 election. Engagement rates increased and covered a widening range of public policy issues, national security issues, and issues pertinent to younger voters [...]. IRA posts on Instagram and Facebook increased substantially after the election, with Instagram seeing the greatest increase in IRA activity.[19]

It is hard to quantify the exact impact the IRA operations had on the election itself, and it would be analytically unsound to argue that the Trump victory is wholly attributable to "Russian interference." Doing so is a convenient excuse to look away from all the other reasons why Americans have elected, and may well re-elect, a populist like Trump to the White House. Nonetheless, it is equally wrong to dismiss these attacks as a hoax or meaningless. They illustrate clearly the ways in which bad actors are attempting to exploit the conditions of the Infocalypse, a trend that will only increase as new developments like deepfakes become more widespread. True to form, Russia continues to exploit the untrustworthy and dangerous aspects of the information ecosystem to find new ways to attack the United States in 2020. Just as Operation Lakhta was more potent than Operation Infektion, Kremlin operations in 2020 are yet more sophisticated still.

OPERATION DOUBLE DECEIT

Ahead of the 2020 presidential race, Russia was working to attack the United States in new ways that made it even harder to trace back its operations to them. There were some clues as to how the Kremlin's tactics were evolving. On 12 March 2020, CNN, Twitter, Facebook and the network analytics firm Graphika[20] exposed Operation Double Deceit. Like Operation Lakhta before it, Operation Double Deceit was a cross-platform social-media influence operation. Only nine months old when it was discovered, the Double

Deceit network was still fairly small. It was composed of 69 Facebook pages with 13,500 followers, 85 Instagram accounts with 263,000 followers, and 71 Twitter accounts with 68,500 followers.

While Project Lakhta was run by IRA agents in St. Petersburg, in 2020 the work had been outsourced to Ghana. In Ghana, the IRA was operating behind the front of an African non-governmental organization (NGO). This fake NGO was called EBLA (Eliminating Barriers to the Liberation of Africa), and it had a website, office buildings and employees. According to a garbled statement on EBLA's (now defunct) fake website, it was using "the cyber activism approach—a mechanism where advocacy is done through the usage of the New Media (NM), to create awareness of human right issues in Africa and beyond via the sharing of stories or news on daily human right abuses."[21]

At first glance, the EBLA website looked legitimate. It referred to an array of "projects" and had a big "donate" button. But, on closer inspection, the façade quickly crumbled. The donate buttons did not work, the site was populated with Latin-stock text, and one of its projects claimed to have raised $231.53 billion—over three times Ghana's annual GDP.

EBLA employees were given smartphones and told to post content on social media. They were asked to post during the late afternoon through the night in local Ghanaian time to correspond with U.S. daytime. It was not difficult for EBLA employees to deduce that they were targeting U.S. audiences, but it was another leap from there to figure out that they were being unwittingly used as agents of the IRA. This double deceit—practised first on Americans, and then even on the Ghanaian operatives—led this operation to be dubbed "Double Deceit" by Graphika. As the analytics firm reports:

> It is unclear how much EBLA's staff knew about the purpose of the operation and the individuals associated with it [...]. EBLA's manager presented himself to them under a false name and told the organization's staff that their job was online activism; an EBLA staff member interviewed by CNN said they had "no idea" they would be working as a Russian troll.[22]

It was not the first time Russia had recruited people to work for them unwittingly. This is something the Kremlin has become very good at in the Infocalypse. In Project Lakhta, the IRA recruited Americans to run, advertise and attend IRA events. Now, it was recruiting Ghanaians to attack America for the Russians.

Double Deceit was again aiming for America's most vulnerable of spots: race relations. EBLA employees were instructed to use the same strategy that had been used in Operation Lakhta. First, they would invite African-Americans to join online communities and try to instil pride in their "tribal" identity using positive messages. Then, they would alienate them from the rest of society, by feeding these manufactured communities with negative messages, making them feel angry and isolated. Their posts combined positive and uplifting messages on issues such as black pride, beauty and heritage with negative and divisive content on racism, oppression and police violence.[23] The Ghanaian operatives were encouraged to play with content to see what would stick and go viral, just as the IRA agents had done years earlier. EBLA employees had access to a shared stock of memes and images to help them. Many were simply recycled from the IRA's stock from 2016.[24]

Project Lakhta may have been exposed and shut down, but Double Deceit shows how Russia is experimenting with new ways to exploit the Infocalypse. Now the IRA was using real people in Africa, it did not need to make up "fake personas" by stealing images and inventing names as it had done in Operation Lakhta.

Double Deceit mixed up pages and communities which

were pretending to be authentically American with pages that were openly located in Ghana. It also contained real social-media profiles of its operatives. This meant that there was a level of "honesty" in some of the content, even though, in sum, this strategy was actually meant to create further confusion. It muddied the trail as to who was ultimately responsible. Double Deceit may have been exposed, but it would be surprising if Russia did not have other pokers in the fire ahead of the election later in 2020.

INCREASINGLY POTENT

Although the basic principles of Russian disinformation are the same, the methods have become much more potent in the Infocalypse. Compare the impact of Operation Infektion (which was already extremely successful) to that of Operation Lakhta, for example. Operation Infektion took the better part of a decade to go global and gather enough momentum to be printed in the newspapers of over 80 countries. This exposed hundreds of thousands of people to this Soviet fabrication. Over the decades, the influence of Operation Infektion did not wane. It grew. To this day, millions of people still believe the AIDS-creation myth.

Project Lakhta went further than promoting a single

lie via foreign print media. Thanks to the Internet, social media and smartphones, the IRA was able to directly infiltrate American public discourse to spread multiple lies with immediacy. The IRA went further by building new divides directly into the DNA of American society, rather than just exploiting existing ones. By the 2010s, multiple lies promoted by the IRA in Project Lakhta went viral on social media within hours or days. Where hundreds of thousands were exposed to Operation Infektion, "tens of millions"[25] of Americans were exposed to Project Lakhta. The legacy of Operation Infektion haunts the United States to this day. Similarly, the effects of the IRA operations in Project Lakhta will continue to grow.

Russia's favoured practice of race baiting is especially dastardly, given the depth of these divides in American society. In the Infocalypse, these tensions quickly spiral out of control. In times of heightened anxiety, mis- and disinformation are dangerous—and deepfakes could be more dangerous still. Released at the right moment, and with the understanding of how to make this content go viral, a deepfake video inciting racial hatred, for example, could cause instant pandemonium.

Recent Russian attacks on the United States continue to exploit divides and embed paranoia, ushering a fraught nation to the brink. While the people of the United States were generally united against the Soviets during the Cold War, that is no longer the case. Remember the words of Yuri Bezmenov: the aim of these attacks is to "change the perception of reality of every American" so "no-one is able to come to sensible conclusions in the interest of defending themselves, their families, their community and their country." It is working.

According to American intelligence services and military, the United States is still ill-equipped to counter the increasingly bold tactics of the Kremlin in the Infocalypse. A Pentagon white paper concluded that "the U.S. is still underestimating the scope of Russia's aggression, which includes the use of propaganda and disinformation to sway public opinion across Europe, Central Asia, Africa and Latin America."[26] Yet, as I cover in Chapter 3, the partisan polarization is so great within the United States that Americans cannot even agree on the objective fact of Russian interference, let alone on *how* it is happening and what to do about it.

The very idea of "Russian interference" is so controversial in political discourse that it is treated as a partisan issue. On the one hand, the Democrats got carried away with theories about collusion between the Trump campaign and the Kremlin, and how it would lead to the certain impeachment of the President. When the Mueller Inquiry finally concluded that there was insufficient evidence to prove beyond a reasonable doubt that Trump's campaign conducted a criminal conspiracy with the Russians, Trump proclaimed that Mueller had exonerated him of "collusion" and of obstruction of justice. Mueller had done no such thing. He later testified to Congress: "We did not address 'collusion,' which is not a legal term. Rather, we focused on whether the evidence was sufficient to charge any member of the campaign with taking part in a criminal conspiracy. It was not."[27] Mueller was careful to point out that the President was not exonerated of wrongdoing, even if a critical threshold for criminal conspiracy was not met. Still, this did not give Democrats the smoking gun they were hoping for. They later went on to persist with an overly politized impeachment process on a separate issue, even though they knew it would fail.

On the other hand, the President of the United States, the supposed first line of defence when it comes to national security, refuses to engage seriously with Russian aggression. Instead, for the last four years, Trump has doubled down on the narrative that the "Russia Hoax" is aimed at de-legitimizing his presidency. He has launched vendettas against anyone, including the U.S. intelligence community, who suggests otherwise. It must be painful for American intelligence chiefs to see their commander-in-chief making a joke of it. When meeting Vladimir Putin at the G20 Summit in Osaka, Japan, in 2019, Trump sarcastically reprimanded a grinning Putin, telling him, "Don't meddle in the election, please."[28] Putin must have loved that.

A WIDER PROBLEM

Russia is inspiring other states, too. If Moscow can get away with these attacks it sends a signal to the rest of the world. The United States and its Western allies appear vulnerable and immobilized in the face of the Infocalypse. The Pentagon found that other state actors, notably China, are sympathetic to Russian strategic aims against the United States and the West, due to their "growing alignment" which sees them "share a fear of the United States' international alliances and an affinity for authoritarian stability."[29]

To understand more about how other rogue and authoritarian state actors are interpreting and emulating Russian actions in the Infocalypse, I spoke to Renée DiResta about the geopolitics of information warfare. Renée told me that Russia is still the master in the field, in part because "it makes much more of a long-term investment." Moscow is prepared to lay the groundwork for operations "over years and decades." I asked Renée about China. She told me that while Beijing is not as strong as Russia on foreign operations, it has the same kind of "long-term game plan domestically." So far, China has successfully exploited the Infocalypse "mostly against its own citizens," having been hugely successful in that endeavour due to the fact that it controls its Internet ecosystem. There, it can monitor and controls billions of Chinese citizens on platforms including Weibo, WeChat and QQ. However, China is increasingly starting to look beyond its own borders. Researchers at the University of Oxford found that the Hong Kong protests in summer 2019 marked a turning point, with China starting to show an "aggressive" new interest in infiltrating Western platforms including Facebook, Twitter and YouTube, something they say should "raise concerns for democracies."[30]

When Renée described some of the Chinese-sponsored operations she has seen on Twitter, it was clear that the Russians are still in a different league. Around the Hong Kong protests, Renée told me she had come across pro-China Twitter accounts which were "sloppy bot-net garbage." Their creators had not even bothered clearing out their past timeline to disguise the fact that they were fake, appropriated accounts. Scrolling down the timeline of past tweets, "you'd find that they'd be tweeting about Ariana

Grande, when suddenly they'd learn to speak Chinese and develop a passionate interest in Hong Kong politics," Renée said, laughing. But don't expect Beijing to remain this clumsy for much longer. As will be explored in Chapter 6, the volatile geopolitical situation around Covid-19 has seen China quickly learn how it can better exploit the Infocalypse.

It is not only China and Russia who are using the chaos of the Infocalypse to their advantage. Renée and I discussed other emerging actors: Iran, Saudi Arabia, the United Arab Emirates and North Korea. Oxford University researchers found evidence of 28 countries practising some form of online disinformation operations in 2017. In 2020, this number has risen to 70. While all these state actors pose serious threats in the Infocalypse, for now there are orders of magnitude between the sophistication of the Kremlin's efforts and those of the others. In one study from 2019, researchers from Princeton found that Russia was responsible for 72 per cent of all foreign disinformation operations between 2013 and 2019.[31] This makes it almost three times as aggressive as all the others combined. If Russia is a maestro playing a Tchaikovsky sonata from memory on a grand piano, then the others are playing "Twinkle, Twinkle, Little Star" on a child's keyboard.

3

The West

the internal threat

President Donald J Trump is briefing the press. The United States, like the rest of the world, is in the throes of its deepest crisis since the Second World War owing to Covid-19. Billions of people are in lockdown and have been instructed to use "social distancing" to fight the virus's spread. With the world economy at stake, scientists are in a race against time to develop a vaccine. The "Leader of the Free World" is asked about possible cures for Covid-19. He is flanked by his Covid-19 taskforce advisor, a public-health expert, Dr. Deborah L Birx. Having glanced at Birx, Trump begins speaking. "So supposing we hit the body with a tremendous—whether it's ultraviolet or just very powerful light," he says, turning to Birx completely, "and I think you said that hasn't been checked but you're going to test it." He swivels back to face the press. "And then I see the disinfectant where it knocks it out in a minute. One minute. And is there a way we can do something like that, by injection inside or almost a cleaning?" He quickly looks at Birx. Then, pointing to his head, Trump continues, "I'm not a doctor. But I'm, like, a person that has a good you-know-what."

This is not fiction. This is the United States of America in 2020. The President is briefing the press that "injecting disinfectant" may cure the virus that has plunged the world into the deepest crisis of this century so far. However, this type of performance is no longer a surprise. Rather, it is par for the course—another normal day of the Trump presidency. The leader of the free world openly disseminates bad information, including outright lies and disinformation. It should not be regarded as partisan to say that while many politicians lie and mislead, Trump outdoes them all. His rejection of the Western Enlightenment ideals of objectivity, reason and truth that underpin liberal democracies is particularly dangerous in the Infocalypse.

How on earth did we get here?

BELLWETHER FOR THE WORLD

Much of the discourse relating to the West's problems with disinformation and misinformation is centred around foreign information operations against the West by the likes of Russia. However, equally important information threats lurk within. These domestic threats are increasingly potent and could yet prove to be the larger existential issue. Nowhere is this more apparent than in the elevation of a man who is perhaps the embodiment of the Infocalypse—Donald J. Trump—to the White House.

Trump rose to power as a populist exploiting the crisis of trust in politics and institutions that has been plaguing the West for years. I have witnessed a similar dynamic unfolding in Europe over the last decade. Much has been written on resurgent populism in the West and the potential implications for liberal democracies, which I will not rehash here. I am particularly interested in one facet of this phenomenon: how populist leaders are normalizing and perpetuating our increasingly corroded information ecosystem, and whether this will lead to a tipping point when Western political systems and society can no longer cope. Will robust political debate and societal progress be possible if our shared sense of reality collapses into a ceaseless domestic information war?

Donald Trump's presidency is a good starting point in trying to answer this question. The manner in which he deploys his populist playbook in this new information ecosystem seems to be bringing the United States ever closer to a tipping point. This manifests in several ways. First, he is making the crisis of trust worse. Secondly, he is dominating the information ecosystem and using his outsized influence to spread copious amounts of bad information, including cheapfakes and deepfakes. And third, he is actively feeding the partisan polarization that reduces everyone's willingness to find common ground. It is no coincidence that that United States has been gripped by societal unrest this year.

Trump is just one actor in the Infocalypse, but I have chosen to focus on him in particular given his importance in defining in what may unfold next. Ultimately, the United States is a bellwether for the rest of the Western world. The manner in which the United States responds (or fails to respond) to its domestic information threat—as personified by its president—will set the tone for the rest of the Western World. Ahead of the 2020 election, the stakes could not be higher.

CRISIS OF TRUST

The Simpsons predicted Donald Trump's ascent to the White House as an exercise in absurdist comedy. Most people did not expect it to happen in real life, until it suddenly did. His victory was no one-off fluke, however. Just as deepfakes did not emerge in a vacuum, neither did Trump. One of the reasons Trump was elected in 2016 was the growing crisis of trust that has come to characterize the politics of the Western world over the last two decades. As I witnessed on the other side of the Atlantic working on Brexit and other European political events, voters increasingly felt that "the system" was no longer "working" for them. They rejected the status quo at the ballot box.

There is no one overarching reason for this growing crisis of trust in Western democracies, but it can be attributed to a variety of factors, including the 2008 financial crash, globalization, immigration, technological change and the Infocalypse. This alienation is evident across the political spectrum, from left to right, and almost universally manifests in anger against the "establishment," "elites" and "media"—all shorthand for the bastions and institutions of the Western democratic world.

Several studies have tracked how this breakdown in trust has accelerated in the last decade, coinciding with the Infocalypse. Take the flagship *Freedom in the World* report, produced by the American NGO Freedom House. Its 2020 edition marks the fourteenth year in which it finds democracy in decline around the world, including in the United States and elsewhere in the West. The report says:

> Democracy and pluralism are under assault.
> Dictators are toiling to stamp out the last vestiges
> of domestic dissent and spread their harmful
> influence to new corners of the world. At the same
> time, many freely elected leaders are dramatically

narrowing their concerns to a blinkered interpretation of the national interest.[1]

I was confronted by this sense of public disillusionment when, in 2018, I helped design the "Democracy Perception Index" with Dalia Research—a public opinion and consumer insights company. We sought to quantify this crisis of trust and conducted one the single biggest surveys of its kind to measure it. We polled approximately 125,000 people in 50 countries around the world, asking them a series of question to gauge how they felt their political "system" was working for them. The data was stark. It showed that almost two-thirds (64 per cent) of people living in democracies felt that their government "rarely" or "never" acts in the interest of the public. This was more than 20 points higher than the 41 per cent of people living in non-democratic countries who said the same. This disillusionment was higher than average in the United States (66 per cent) and the UK (65 per cent). In France and Germany, it was 64 per cent.[2]

This crisis of trust that has come to define the Western world has served Trump well. During his candidacy, his detractors scoffed that he was unfit for high office, not least because he so openly and flagrantly lied. He was one of the most vocal advocates of the "Birther Conspiracy"—that

President Barack Obama was not actually born a U.S. citizen.[3] He called climate change a "big hoax" to "benefit China." He implied that the father of Republican Senator Ted Cruz might have had something to do with the murder of John F Kennedy. To the amazement of his critics, Trump won regardless. This was not because his supporters were "too stupid" to realize that he lies. It is widely understood that he lies. According to one 2018 study, fewer than three in ten Americans— including fewer than four in ten Republicans—believed several prominent false claims by Trump.[4] To paraphrase the German-American philosopher Hannah Arendt, it doesn't matter that leaders lie if people think that everything is already a lie anyway. For the American citizens who felt such intense alienation from "the system," it might feel as though Trump lies for "the right reasons," or that he "lies for them." If everyone is a hypocrite, why bother with the truth?

Since his election, Trump has not modified his behaviour to remedy this disillusionment. If anything, he has further contaminated American political discourse by actively spreading mis- and disinformation. Objectively, he is a lying machine. According to a database set up by *The Washington Post*, Trump had made over 18,000 false and misleading claims in the three years up to January 2020. That is an average of 15 false claims a day from the President of the United States. This is accompanied by daily attacks on American institutions, the pillars of its democracy—be that the intelligence community, the media or even the government itself.

Aside from lying, Trump is also adept at utilizing the "liar's dividend." The liar's dividend is the concept of a liar dismissing anything they don't like as "fake," even if it is not. The liar's dividend is becoming an ever more powerful tool in the Infocalypse. Trump uses it constantly, for example in his daily refrains of "fake news" against information he does not like. He even uses it against video evidence. In 2017 he started dismissing the infamous *Access Hollywood* tape in which he bragged of grabbing women "by the pussy" as "fake." (Of course, when deepfakes become more widespread, the liar's dividend will become even more powerful. In a world where anyone can be faked, everyone has plausible deniability.)

In short, it is fair to conclude that Trump is perpetuating the very crisis of trust that helped elect him. The Pew Research Center, a non-partisan U.S. think tank based in Washington, DC, has documented how trust in the government has been on a downward spiral in the United States since 2007—when for the first time, only 30 per cent of the American public said that they could trust the government to do what is right. Under Trump, it has sunk to historic lows. According to Pew, in 2020, only 17 per cent of Americans said they could trust the government in Washington to do what was right "just about always" (3 per cent) or "most of the time" (14 per cent).[5] This data suggests that the United States may be moving closer to the "tipping point."

AGENDA SETTING

A second indicator that this "tipping point" is fast approaching is the fact that the president actively spreads so much bad information. Trump dominates the information space by "agenda setting." A theory of power, this suggests that the more something comes up in the public debate, the more the nation thinks it important. By this logic, the media, the conduit for information, has the power to literally "set the agenda." From the very beginning of his candidacy in 2015, Trump has turned the idea of agenda setting on its head. Instead of letting the media set the agenda, *he* sets the agenda by dominating traditional media. One way he does this is by creating scenarios where the media simply cannot resist covering him. From the moment he launched his campaign for President, when he rode down the escalators at Trump Tower, flanked by the photogenic members of "Trump Inc.," calling Mexicans "rapists"[6] and promising to "beat China," he had us—hook, line and sinker.

His grip on the news cycle allows him to flood it with as much mis- and disinformation as he pleases. For example, no longer able to travel freely and hold rallies during the Covid-19 pandemic, Trump turned his daily coronavirus press briefing into an agenda-setting masterclass. Objectively, his response to the crisis was abysmal (see Chapter 6 for more). When the President talks about injecting disinfectant as a possible treatment for the novel coronavirus, it is surely not a partisan statement to say his behaviour is incompetent, raising serious questions over whether he is qualified to be President. Yet Trump uses press conferences to great effect to spin his "alternative facts," including patently untrue claims that he was "very fast" to react to the pandemic.

Polls show that American attitudes towards the virus are linked by partisan political beliefs. In a public-health crisis, this has potentially dangerous consequences. As the Covid-19 death toll in the United States passes 100,000 at the time of writing, it is possible that Americans will not take adequate measures to protect themselves based on flawed information shaped by partisan political messaging—by flouting lockdown, for example. One study by the Kaiser Family Foundation found that only 37 per cent of Republicans (versus 70 per cent of Democrats) said they wore a mask when leaving the house. The same study suggested that Trump may not be punished at the ballot box for his bad handling of Coronavirus. Only 6 per cent of Republicans said that the pandemic was a voting priority, compared to 29 per cent of Democrats.[7]

Commentators lamenting Trump's lies around Covid-19 make an important point. But they also miss an important one. It is the volume of coverage—good or bad—that lets Trump do what he does best: set the agenda. As the old saying goes, "All press is good press." Trump agrees. He tweeted a quote from a *New York Times* article to boast about just how much coverage he is getting.

> **President Trump is a ratings hit. Since reviving the daily White House briefing, Mr. Trump and his coronavirus updates have attracted an average audience of 8.5 million on cable news, roughly the viewership of the season finale of "The Bachelor." Numbers are continuing to rise ...[8]**

Trump sets the agenda through social media too. The importance of Twitter as a tool for the President cannot be understated. It provides him with a soapbox that bypasses traditional media gatekeepers and allows him to communicate directly to the world. In the summer of 2016, Trump already had a substantial Twitter following of 10.2 million, tweeting an average of 15 times per day.[9] Far from scaling back his tweeting when he became Commander in Chief, Trump doubled down. At time of writing he has 80 million followers— a number that has jumped by a few million since the start of 2020. He is almost twice as prolific as he was four years ago: he now averages 28 tweets a day.

Every day, Trump's Twitter feed is a non-stop frenzy of activity. Anyone from Kentucky to Khartoum can follow what Trump is saying or retweeting on Twitter with almost hourly updates. His tweets and retweets are incendiary, outrageous and bold. They are also often straight-up lies. There is, of course, a large element of showmanship to all of this. For example, when he proclaimed himself "a very stable genius!"[10] on Twitter, or when he attacked the Iranian president Hassan Rouhani with:

> "NEVER, EVER THREATEN THE UNITED STATES AGAIN OR YOU WILL SUFFER CONSEQUENCES THE LIKES OF WHICH FEW THROUGHOUT HISTORY HAVE EVER SUFFERED BEFORE. WE ARE NO LONGER A COUNTRY THAT WILL STAND FOR YOUR DEMENTED WORDS OF VIOLENCE & DEATH. BE CAUTIOUS!"[11]

This "entertainment" factor, which somehow disguises the dangerous nature of the content, also helps normalize it. Everything Trump tweets goes viral. He reaches tens of millions of people around the world in a way that no leader in history has ever done before. Trump's sheer domination of the information space can also be interpreted as a type of censorship. A tactic academics call "censorship through noise," it is a classic tool of disinformation. By flooding the zone with so much information that no-one can keep up, confusion and distraction ensue. Like the Russians, Trump excels at this tactic.

CHEAPFAKES AND DEEPFAKES

Another indicator of the tipping point is the fact that the President of the United States has demonstrated that he is willing to openly share fake and manipulated audiovisual materials. Since 2018, U.S. political discourse has been peppered with cheapfakes. The forebear of deepfakes, these are mis-contextualized or crudely edited pieces of video, audio or images. The first notable cheapfakes incident occurred in November 2018, right after the U.S. midterm elections. The (then) CNN White House correspondent, Jim Acosta, someone who had a long history of sparring with

Trump, was involved in a testy exchange with the President at a press conference. When a young White House intern tried to grab the mic away from him, Acosta lowered his arm to shield the roving mic from the woman, saying, "Excuse me, ma'am," and continued his questioning of the President. Trump went ballistic. In an unprecedented reprimand, Acosta's "hard pass," which allowed him White House access, was suspended "until further notice." The White House Press Secretary, Sarah Sanders, justified the decision by accusing Acosta of having "put his hands on a young woman just trying to do her job."[12]

The next day, a doctored video of the exchange was circulated by the far-right InfoWars website. It had been manipulated so that it looked as if Acosta was meting out a blow on the intern with his arm. This supported the White House version of events, and the White House retweeted the doctored video, justifying their decision to suspend Acosta. When it became apparent that the video was manipulated—and that therefore there was no reason to suspend Acosta—the White House doubled down, saying that Acosta's general behaviour was inappropriate, and that it was in the power of the White House to decide who got access. It was only when CNN looked to take the case to court that the White House backed down, restoring Acosta's pass.[13]

The Acosta incident was just the first of several other doctored and manipulated videos that have since gone viral. In 2019, there was a video that was edited to make it look as if Nancy Pelosi, the Democrat and U.S. House Speaker, was slurring her words, appearing to be drunk. Digital forensics experts say the video was probably slowed down to achieve this effect. The President seized upon the cheapfake when it came to his attention. He retweeted it gleefully, amplifying it to his millions of followers and using it to push the message that "Nervous Nancy" Pelosi was mentally impaired.

Since then, cheapfake video content has become a regular feature of Trump's output on Twitter. In February, he capitalized on a cheapfake that was created off the back of a political stunt by Pelosi. While Trump was delivering his State of the Union speech, Pelosi took the dramatic and politicized step of ripping up a copy of the President's speech while he was speaking. Later, an edited video emerged that made it look as if Pelosi was ripping up the speech in response to individual audience stories, including that of a former Tuskegee Airman (a Second World War veteran) and a single mother whose daughter received a scholarship that night.[14] While Pelosi's act was no doubt hyper-partisan, it was done in defiance of the President, rather than to pour scorn on American citizens (as the cheapfake Trump retweeted implied).

Since then, Trump has also retweeted a deceptively edited video of Joe Biden in which he seems to back Trump for president by appearing to say, "We can only re-elect Donald Trump."[15] The video first appeared on the feed of Dan Scavino, Trump's social-media director. With the platforms increasingly coming under pressure to monitor disinformation ahead of the election, Scavino's tweet was the first to be labelled with a health warning under Twitter's new rules for manipulated media. That did not stop Trump, who retweeted the cheapfake with the caption, "I agree with Joe!"

Predictably, Trump has already graduated from these cheapfakes to his first deepfake.

Again, it was Biden in the firing line. In April 2020, Donald Trump retweeted a deepfake of his presumptive Democratic election opponent. It was a GIF generated using an AI-powered smartphone app called Mug Life. In the animation, Biden's hands are clasped together in front of him. He raises his eyebrows and gurns as he rolls his tongue round the outside of his mouth, licking his own face.[16] The tweet originated on an anonymous Twitter account, @SilERabbit, which claims to be a generic parody account but is full of anti-Biden content.

The animation is marked by a Mug Life watermark in the bottom right-hand corner, so it is clear that it is a deepfake.

But the intent behind it is malicious. The GIF is accompanied by a caption calling Biden "Sloppy Joe" and the animation makes Biden look stupid. @SilERabbit had tweeted the animation of Biden several times before hitting the jackpot on 27 April. Having never engaged with @SilERabbit before, Trump suddenly retweeted the tweet. It immediately went viral. Soon, the hashtag "Sloppy Joe" started trending on Twitter.

Was the animation silly? Yes. Was it clearly fake? Yes. But was it effective? Absolutely. Since Trump amplified this tweet, it has had almost 17,000 retweets, 40,000 "likes" and has been seen by further thousands. It is reasonable to conclude that it may have had at least some impact on how voters might view Biden. I have already covered how memes and other "silly" pieces of content are a pernicious way of spreading political disinformation, precisely because they seem so innocent. Someone hiding behind a supposed parody account playing off the moniker "Silly Rabbit" (@SilERabbit) can still do damage, especially when amplified by an "influencer" like Trump. In this case, it just so happens that the influencer is the President, normalizing the use of manipulated media to attack political opponents.

PARTISAN POLARIZATION

Trump is also bringing the United States closer to the tipping point by using information warfare to entrench partisan polarization into American society. While reinforcing the partisan divide suits Trump's political objectives, it makes the United States (and the West generally) weaker. If Americans are too busy fighting one another, they cannot respond to existential threats that requite bi-partisan solutions (like the Infocalypse).

American politics have always been combative, but the extreme partisan polarization that has emerged in the Infocalypse is a relatively new phenomenon. This trend pre-dates Trump, but it has been accelerating under him. Data from the Pew Research Center shows that it started forming as a big dividing line in American society in 2012. Under Trump, it has become *the* biggest dividing line."[17] Today, 91 per cent (!) of Americans think there are "strong" or "very strong" conflicts between the Republicans and the Democrats. Only 59 per cent say the same about the conflict between rich and poor, and 53 per cent about the conflict between black and white. Political allegiance is a bigger source of division than race or class.

To understand a bit more about how Trump actively drives this partisan polarization, I spoke to Dr. Jennifer

Mercieca, a historian of political rhetoric at Texas A&M University, who has spent several years studying the way Trump communicates.[18] "A lot of people don't like to hear it," says Jennifer, but "Trump is a rhetorical genius." He uses his rhetoric as a political tool. On the one hand it binds his followers to him, and on the other it denigrates his detractors. Given his domination of the information ecosystem, his words are having an outsized influence in entrenching the growing polarization in the United States.

To bind his followers to him, Trump employs three main rhetorical devices. First, he makes "appeals to the crowd." Trump uses this device to praise his supporters for backing him. Often his words have no basis in reality. "We're winning everywhere," for example, and "I think we've done more than perhaps any President."[19] He routinely makes such false claims about the U.S. economy, including "We had probably the best economy in the history of the world, bigger than China, bigger than anybody." He is doing this even during the current Covid-19 recession (which has seen Americans file 40 million jobless claims.)

Secondly, Trump artfully deploys *paralipsis,* a rhetorical ploy that allows him to say something by denying that it is being said. In Jennifer's words, this allows Trump to "recirculate information without being held accountable for it." An example would be when Trump insulted the North Korean dictator by denying that he would insult him, tweeting "Why would Kim Jong-un insult me by calling me 'old,' when I would never call him 'short and fat'?" When Trump uses paralipsis to say things like "I'm not gonna tell ya..." but then proceeds to do just that, it makes his followers feel that Trump is revealing the awful truth, even if what follows is a complete fabrication.

Thirdly, Trump appropriates the narrative of American exceptionalism to serve his own objectives. This narrative of the United States as different and special has a long history in American political discourse. Trump's version is novel in that it centres around claims which are often (but not always) misleading or false about how badly the United States has been treated by others, and how only he and his supporters can restore the United States to its former glory. To "Make America Great Again." This is commonly seen in Trump's communication on foreign policy, where he makes statements like "China never paid us 10 cents! China has taken advantage of the United States—until I came here!"[20] or "Nobody in 50 years has been WEAKER on China than Sleepy Joe Biden. He was asleep at the wheel. He gave them EVERYTHING they wanted, including rip-off Trade Deals. I am getting it all back!"[21]

Aside from building up his own followers, Trump also tears down his detractors. Again, he uses three primary rhetorical devices. First, he advances conspiracy theories to make threats. For example, he whipped up rallies of thousands of people to chant, "Lock her up! Lock her up! Lock her up!" on spurious allegations that his 2016 adversary, Hillary Clinton, was a "criminal." In 2020, Trump actively started amplifying a conspiracy theory known as "Obamagate," which is centred around false allegations that Barack Obama illegally sabotaged Trump's presidency. While it is not clear exactly what Trump is accusing Obama of, he says that Obama is guilty of "the biggest political crime in American history, by far!" In May 2020, he repeatedly referenced "OBAMAGATE!" in his tweets, claiming that it "MAKES WATERGATE LOOK LIKE SMALL POTATOES!"[22] References to this supposed conspiracy are gathering speed ahead of the election.

Secondly, Trump is a master of *ad hominem* attacks, which means he attacks the player rather than the ball. If asked a question he does not like, he deflects by attacking the person asking it. When Peter Alexander, the NBC White House correspondent, asked the U.S. President what he would say to Americans who are scared in the current pandemic, Trump unleashed a tirade. He snapped back, "I say that you're a terrible reporter! That's what I say. I think it's a very nasty question and I think it's a very bad signal that you're putting out to the American people."[23] (The exchange made headlines around the world—but rather than focusing on the administration's lack of preparedness, the story became about the *ad hominem* attack. Advantage: Trump.) Trump also uses catchy nicknames that help his *ad hominem* attacks stick in the public imagination—"Crooked Hillary" (Clinton), "Sleepy Joe" (Biden), "Nervous Nancy" (Pelosi) and "Cheatin' Obama." He does the same to organizations and institutions, like the "Failing New York Times," the " Do Nothing Dems" and the "Deep State Department" (that is to say, the State Department).

Thirdly, Trump also relies on a device called reification— speaking of others like objects—to denigrate them as enemies. As Jennifer Mercierca explained, this type of dehumanization is "a war tactic." When you refer to people as vermin, viruses or scum, "you are preparing to do battle with them." Mexicans are "rapists and murderers," but Trump not only uses this kind of language against illegal immigrants (as you might expect him to) but even against the small group of "Never Trump" Republicans. As he tweeted in 2019, "The Never Trumper Republicans, though on respirators with not many left, are in certain ways worse and more dangerous for our Country than the Do Nothing Democrats. Watch out for them, they are human scum!'[24]

Trump's populist approach in the Infocalypse is making the United States a more divided and dangerous country. The distrust and polarization he normalizes in this increasingly corroded information ecosystem means that it will not take much for violence to spill over into real life. In the Infocalypse, that violence can spread faster and become ever more difficult to control: another sign that the United States is creeping closer to a dangerous tipping point.

"I CAN'T BREATHE"

It started just as this book was about to go to press. On 25 May 2020, a 46-year-old black man, George Floyd, was killed by a white police officer, Derek Chauvin, in Minneapolis, Minnesota. Another senseless loss of black life due to police brutality. Floyd was arrested for allegedly using a counterfeit $20 bill. In a horrific video captured by a passer-by, Chauvin suffocates Floyd by kneeling on his neck for almost nine minutes, while Floyd begs for his life. The county medical examiner later declared it a homicide.[25] Floyd's final pleas are more powerful than anything I can write about this grievous injustice:

I cannot breathe

I cannot breathe

they gon' kill me

they gon' kill me

I can't breathe

I can't breathe

please sir

please

please

please I can't breathe

The phrase "I can't breathe" is already synonymous with race relations in the United States. In 2014, another black American, Eric Garner, used the same words when he was put in a chokehold and killed by a white police officer.[26] In the context of increasing racial tension and grievances with police brutality, Floyd's killing was the straw that broke the camel's back. Within hours of the video appearing, protest and violence broke out. Starting in Minneapolis, within days it had spread to 30 cities across the United States. In some places, legitimate peaceful protest was hijacked by angry mobs and anarchists. Some began looting and setting buildings, businesses and cars on fire. Meanwhile, police and armed forces responded with tear gas and rubber bullets, and cities instituted curfews and states of emergency. Some cities even called in the National Guard to restore order. Online and on the ground, the flames of anger were fanned both by white supremacists and by left-wing extremists.[27]

As civil order threatened to crumble entirel, President Trump unconscionably and actively made the situation worse. Rather than acknowledging the peaceful protesters, he dismissed them as members of Antifa (a militant left-wing group) and as the "Radical Left."

As violent protest broke out in Minneapolis, the President blamed the city's Democratic mayor, Jacob Frey, tweeting "How come all of these places that are defended so poorly are run by Liberal Democrats?"[28] After one man was shot and killed on another day of violence in Minneapolis, Trump tweeted the phrase "When the looting starts, the shooting starts"—a racially loaded expression dating back to the 1960s.[29] Then, when angry protestors gathered outside the White House, Trump sent off a barrage of tweets threatening that if they came any closer they would be "greeted with the most vicious dogs, and most ominous weapons, I have ever seen."[30]

As journalists took to the streets to cover what was unfolding, many were assaulted and arrested amid the chaos. Again, instead of calling for calm, Trump doubled down on his attacks against the press, tweeting "Fake News is the

Enemy of the People!"[31] and accusing the press of "spreading more 'disinformation' than any foreign country, including Russia."[32]

In his national address from the White House Rose Garden on 1 June 2020, Trump's rhetoric was more divisive than ever. Instead of appealing for unity, acknowledging the peaceful protestors and condemning all violence, he threatened to invoke the Insurrection Act, which would give him the power to overrule state governors and send in the National Guard, to "dominate the streets."[33] Several Congressional Democrats decried his behaviour as that of a would-be authoritarian leader. Meanwhile, Republicans commended him for restoring law and order.[34] By responding to this evolving situation by actively driving polarization, Trump made something that was about racial injustice into a partisan issue.

Months before the election, the United States found itself stumbling from one national crisis to the next. First, the Covid-19 pandemic, then the subsequent economic crisis, and now finally violence and protest in the wake of the George Floyd killing. Each of these events is dangerous as they are occuring in an information environment which is polarizing and susceptible to mis- and disinformation. And yet, rather than calling for calm, the President continues to pollute the information ecosystem by filling it with rhetoric to feed distrust and polarization. Trump's words and actions matter. In the Infocalypse, bad and divisive information is leading to violence that spreads further faster. Sadly, it appears that more lives will be lost before the United States can come together.

2020 AND THE INFOCALYPSE

Against the backdrop of this brewing storm, Americans will head to the polls in November 2020. That election marks an important moment to take stock of how far the Infocalypse has evolved since the last election. I believe the corrupt information ecosystem will play an even greater role in the 2020 election than it did in 2016, and I have four predictions for how that may play out.

First, there will be more foreign interference. Primarily, it is likely to come from Russia, which is still the best at it. There is no chance that the Kremlin is not seeking to exploit this election. As I covered in Chapter 2, Russia is up to its old tricks, and is cooking up new ways to disguise its actions. While Russia is still the master, others are increasingly inspired by the Kremlin's tactics. This includes Iran and Saudi Arabia, but notably also China. As I cover in more detail in Chapter 6, the deteriorating relationship between Beijing and Washington is leading the former to adopt new aggressive foreign disinformation tactics. Until 2019, China primarily used disinformation at home, but now it is looking to infiltrate Western public life too.

Secondly, domestic disinformation will be rife. Though I have focused on Trump in this chapter given his singular importance, disinformation will come from politically motivated groups and individuals across the political spectrum. We saw indications of this as early as the 2017 special Senate election in Alabama (a Republican stronghold), when Democratic operatives ran "Operation Birmingham." It was a Russian-style disinformation campaign against Republican candidate Roy Moore that used social media to divide Republican Alabamians by encouraging them to vote for a rival Republican candidate through a write-in campaign. It also orchestrated an elaborate "false flag" operation that planted the idea that the Moore campaign was being supported by inauthentic Russian accounts (so-called "bots") on Twitter. In a document later obtained by *The Washington Post*, Operation Birmingham operatives claimed that they had provided the decisive margin in an election in which Moore lost by some 22,000 votes.[35] (While this claim is impossible to verify, it should be noted that there is no evidence that the winner, Democrat Doug Jones, knew anything of Operation Birmingham; indeed, he later called for a federal investigation).

It is an indicator of the future direction of travel. I thus expect to see more such operations in the forthcoming election. Indeed, NewsGuard, an organization that monitors and protects from disinformation, has already released a report on how major Republican and Democratic campaign organizations are using "shady news sites" to spread political propaganda and disinformation.[36] In an increasingly partisan political environment, the United States is facing a race to the bottom when it comes to home-grown disinformation. In an ever-destructive cycle, this in turn will further feed partisan polarization and distrust, making the Infocalypse even worse.

Thirdly, the President will be one of the chief perpetrators of this domestic threat. The sheer magnitude of his influence combined with the scale of his disinformation efforts is potentially more harmful to the United States than any foreign operations. In the context of 2020, he is spreading disinformation aimed at invalidating the election result even before it is held. Trump is openly doubling down on claims that the Democrats are going to organize voter fraud in a "rigged election." As he wrote in a tweet on 26 May 2020:

There is NO WAY (ZERO!) that Mail-In Ballots will be anything less than substantially fraudulent.

Mailboxes will be robbed, ballots will be forged
& even illegally printed out & fraudulently signed.
The Governor of California is sending Ballots
to millions of people anyone [...] living in the
state, no matter who they are or how they got
there, will get one. That will be followed up with
professionals telling all of these people, many of
whom have never even thought of voting before,
how, and for whom, to vote. This will be a
Rigged Election. No way!

Trump's claims of a "rigged election" are now being
repeated almost daily. This is an alarming development, and
it raises the question of what will happen if Trump does not
win in November. So far (and there are some months to go,
so this does not necessarily mean anything), the polls suggest
that Biden is ahead of Trump.[37] Will Trump accept the result
of the election if he loses? Even if he does, will his supporters?
On the flip side, given his behaviour, what will happen if
Trump wins? In the environment of the Infocalypse, it would
not take much for real-life violence to erupt in the aftermath
of the election (consider the George Floyd protests), and it
could be sparked on either side of the political spectrum
depending on who wins.

This brings me to my fourth prediction, and that is that the Infocalypse will continue to evolve, regardless of the election outcome. That is ultimately the real challenge. In this chapter, I have told the story of the Infocalypse through one of its most dangerous actors, Trump, but ultimately, the Infocalypse is bigger than Trump. If Trump left office tomorrow, the Infocalypse would not leave with him. If he has another four years, he will undoubtedly sustain and perpetuate it. All indications suggest that the Infocalypse—the very information system within which we exist—is evolving into an ever-greater threat. Unfortunately, the current state of American politics makes it difficult to see how the bipartisan effort that is needed to ward off the worst effects of the Infocalypse can be mustered. And that is the real tragedy. If a whole-society mobilization against this corrupt and dangerous information ecosystem fails, we all lose.

4

The Rest

global information
disorder

A GLOBAL THREAT

The Infocalypse rages not only in the West, but in the non-Western world—a term by which I am broadly referring to Latin America, Asia and Africa. No political system or country is immune. Mis- and disinformation have been causing chaos and even inciting genocidal violence in places like the Philippines, Myanmar and India—before the West woke up to the problem with the U.S. election of 2016. Outside the West, certain countries are arguably even more vulnerable to the dangers of the Infocalypse. In authoritarian, rogue or unstable regimes, bad actors are able to exploit the chaos of the Infocalypse with greater impunity. Western democracies still have defences in the rule of law, the free press and the democratic institutions, all of which are established (even as they become increasingly vulnerable). In countries where there are no—or fewer—institutional safeguards, however, the consequences of the corroding information ecosystem could be even more devastating. This may be especially harmful to those global citizens (the majority of whom reside in Africa and Asia) who are just about to join this corroding information ecosystem without any strong protections.

Outside the West, the Infocalypse is broadly being used to threaten and intimidate domestic opposition, drown out dissenting opinions, incite violence (ethnic and/or gender) and suppress fundamental human rights. Take the example of Myanmar. When the ruling military junta started to relax its extreme censorship in 2010, the South-East Asian country went from a nation of limited information to one of information abundance almost overnight. As citizens acquired smartphones, millions joined Facebook. For many, Facebook became synonymous with the Internet. By 2019 an estimated 20 million of Myanmar's 53 million citizens had joined the platform.[1] While these freedoms were liberating, they came with new dangers. As citizens were thrown into the wild west of the Infocalypse with no protection, Facebook became a breeding ground for disinformation. In Myanmar, it was used to exacerbate the darkest side of pre-existing communal tensions between the (majority) Buddhist and (minority) Muslim populations. This racial hatred—stirred up online by Buddhist extremists, military leaders and militia—soon spilled over into real life.

In 2014, false accusations that the Muslim owner of a teashop in Mandalay had raped a Buddhist employee were shared by Wirathu, an ultra-nationalist Buddhist monk, on Facebook. Soon, enraged mobs armed with machetes and sticks began marauding around Mandalay, torching cars and ransacking shops. The violence was contained, but the incident was the first warning shot of what was to come.[2] As the racial and ethnic disinformation continued to spread on Facebook, the animosity reached breaking point in 2015, culminating in a monumental horror against the Muslim Rohingya people of Myanmar's Rakhine State. First, they were forcibly displaced. Then, the Burmese army started to carry out a large-scale campaign of ethnic cleansing against them. Human Rights Watch calls the atrocities against the Rohingya "crimes against humanity."[3] By the time Facebook finally started to ban the extremists encouraging violence against the Muslim minority on its platform with mis- and disinformation, 25,000 Rohingya had been murdered and a further 700,000 had fled the country. The UN lambasted Facebook, saying it had "turned into a beast," and that its conduct—allowing disinformation and hate speech against the Rohingya to proliferate—"bears the hallmarks of genocide."[4]

The example of Myanmar is a reminder that the freedoms of the Information Age are not inherently good; indeed, that they can be downright deadly if citizens are exposed to a dangerous information ecosystem.

India provides another example of how dangerous this information ecosystem can be. Take the Facebook-owned messaging app, WhatsApp. The closed and encrypted nature of WhatsApp means that it has been one of the main conduits for the spread of disinformation. Because of its closed nature, it is difficult to identify and trace how this happens, but the impact is huge. WhatsApp has over 2 billion users worldwide, and given that it is primarily used to communicate with friends and family, people are more likely to trust the information they receive on the app. Some 400 million people are now on WhatsApp in India, where it has become a major source of mis- and disinformation. Because millions of Indians are new to the Infocalypse and have barely any protections from it, its dangers have quickly been revealed. There have already been dozens of horrific cases of killings—mostly in rural areas—due to disinformation spreading like wildfire on people's smartphones. Take for example the July 2018 case of five Indian friends who were visiting relatives in a countryside village in South India. On the way there, they stopped for a break near a school—children were just flooding out of the gates on their way home. The men had brought gifts for their relatives, including chocolates, some of which they distributed to the children. What was meant to be an idyllic vacation soon turned to horror. India was on high alert due to rumours of child kidnappers, which had been fuelled for months by stories being circulated on WhatsApp. One piece of viral disinformation included a video in which children playing on the street are approached by two men on a motorbike, one of whom then leans over and grabs a child, before zooming off.

The video was a cheapfake—a mis-contextualized video taken from a child-safety campaign in Pakistan—which drummed up mass hysteria that India was full of child-kidnapping thugs. In the July 2018 case, angry villagers mistook the five innocent vacationers as a threat, approached the cars and started to deflate the tyres, beating the men and accusing them of trying to kidnap the children. Terrified that they would be killed, three of the men managed to escape in the car while the other two ran away through the fields. Meanwhile, a video of the men fleeing in their car had been made and posted on WhatsApp with an accusation that they were child kidnappers. Within minutes it went viral, reaching other villages in the surrounding area. As the car with the three fleeing men approached the next village, the villagers blocked off the road and an angry mob descended onto the car, beating the men viciously. Police were called to the scene, but could not control the mob. One of the men, a 32-year-old software engineer, was brutally beaten to death and the other two were beaten to within an inch of their lives.[5]

DEEPFAKES AND HUMAN RIGHTS

Sam Gregory is programme director at WITNESS, a human-rights organization that has been helping citizens around the world use video technology as evidence to advance human rights. Of course, this mission is predicated on the understanding that video is seen as an incorruptible form of evidence. Deepfakes will turn this on its head. So, Sam has spent a lot of time thinking about how deepfakes will impact on already vulnerable communities and people outside the Western world.

In 2019, WITNESS led a series of workshops with key stakeholders (journalists, activists, NGOs) in Brazil, South Africa and Malaysia, to hear how those who fight what Sam calls "firehoses of falsehoods" in these parts of the world perceived the deepfake threat.[6] Their message was consistent: rather than being worried about foreign interference (as is often the concern in the West), it was their own government, or other domestic forces, that they were mostly concerned about. As WITNESS reports:

> For citizens of these smaller, less geopolitically significant countries, nation-state attacks were not on the radar—or rather were deprioritized in the face of other threats. Instead the offensive

intelligence capabilities of the state were seen as much more likely to be trained on internal voices of opposition, and citizens engaged in political criticism felt that they had more to fear from their own governments than foreign actors.[7]

There is already a very fragile consensus around the idea of citizens' audiovisual evidence in many of these places, and if the government is a despotic or authoritarian regime, there is a real incentive to undermine citizens' media, so that only those in power can control the narrative.

The only widespread use of deepfake video right now—non-consensual pornography—has already been exploited as a way to silence and intimidate women. An investigative journalist and writer, Rana Ayyub is a Muslim woman working in India. Much of her work focuses on the ugly sectarian violence that plagues South Asia as Muslims and Hindus live together in disharmony. She has written a book, for example, on how politicians and police were complicit in the 2002 Gujarat riots,[8] in which 790 Muslims and 253 Hindus were massacred in a pogrom by rampaging mobs. The horrific events are seared into India's collective memory. As the author Pankaj Mishra later wrote, "The pogrom was extensively televised by India's innumerable [...] TV channels.

Many middle-class Indians were shocked to hear how even the very young had not been spared—the slayers of Muslims were seen smashing the heads of children against rocks."[9]

The chief minister of Gujarat at the time was Narendra Modi, later accused of having instructed civil servants and police not to stand in the killers' way. Modi, now serving as the Indian Prime Minister and head of the ruling Hindu nationalist Bharatiya Janata Party (BJP), has always denied involvement and condemned the riots. (He was later cleared of complicity in the violence by a Special Investigation Team appointed by the Supreme Court of India.)

Rana saw things a little differently. As an open and fierce critic of the BJP, she had a track record of digging in the most dangerous of places. This clearly made her a target for online intimidation. Rana told the *Huffington Post*, "I always tried to ignore it by telling myself it's only online hate and it would never translate into offline abuse."[10]

That changed dramatically in April 2018. India was outraged by the case of an eight-year-old Muslim girl who had been raped. Members of the ruling BJP were preparing a march to support the Hindu man accused of the heinous act. Rana was due to appear on the BBC and Al Jazeera, to speak about, to use her own words, "how India was bringing shame on itself by protecting child sex abusers." The next day,

Rana found herself in the middle of a targeted disinformation campaign.

First, a series of fake tweets, purportedly coming from Rana, started circulating on social media. Playing on the Hindu/Muslim divide between Indian and Pakistan, the screenshots of the fake tweets, which looked as if they were coming from Rana's official Twitter account, read "I hate India," "I hate Indians" and "I love Pakistan." Rana had to quickly clarify that the tweets were fake. But the attack escalated. The next day, Rana got a tip-off from a source inside the BJP warning there was a video of her circulating on WhatsApp. He sent it to her. She opened it and started vomiting. It was a fake porn video, and she was the "star." As she later recounted, "I just didn't know what to do. In a country like India, I knew this was a big deal. I didn't know how to react, I just started crying."[11] Her phone started going crazy. Ping, ping, ping—hundreds of Twitter, Facebook and Instagram notifications filled up her screen. She received lurid private messages commenting on her "body" while public comments on her social-media accounts filled up with screenshots of the video. Then, it was shared by the fan page for the BJP. It went viral.

The next day, Rana was "doxxed," meaning her private information (in this case her personal number) was published on the Internet with malicious intent. It was published alongside screenshots of the deepfake porn. Soon, her WhatsApp was full of messages with people asking her for her rates for sex, threatening her with rape and death. As a result, Rana was unable to leave her home for days. She stopped writing. She says, "The effects have stayed with me. From the day the video was published, I have not been the same person. I used to be very opinionated, now I'm much more cautious about what I post online. I've self-censored quite a bit out of necessity."[12]

If the video was released with the aim of silencing her, it hit the bullseye. This is only the very first of many deepfakes that will be used to silence political opposition.

The BJP is also behind the first use of a deepfake video in a political campaign, which emerged in India in February 2020. Vice broke the story to the world, reporting that on the day before the Assembly elections in Delhi, two videos of the BJP President, Manoj Tiwari, criticizing the incumbent Delhi government of Arvind Kejriwal, leader of the left-populist Aam Aadmi Party (AAP), went viral on WhatsApp. While one video had Tiwari speak in English, the other was him speaking in the Hindi dialect of Haryanvi. "[Kejriwal] cheated us on the basis of promises. But now Delhi has a chance to change it all!" said the BJP politician, urging voters to make it right at the ballot box. These videos were then distributed across 5,700 WhatsApp groups in Delhi and surrounding areas, reaching around 15 million people. The video was a deepfake—and it went viral with no indication at all that it was synthetic media.

The Indian political communications firm The Ideaz Factory, which made the video with the BJP, hailed it as an innovative way to use deepfakes for "positive campaigns," by helping politicians connect with a diverse array of voters. As soon as I tweeted a link to the Vice article, criticizing the fact that the video was shared without a clear indication that it was a deepfake, I got a message in my Twitter inbox from Sagar Vishnoi, who works at The Ideaz Factory. "Hello Nina, Greetings," it started. "We did positive deepfake not to spread misinformation or fake news but just to spread the leader's message in different languages. We understand the sensitivity of its negative use."

Given my upbringing in South Asia and my awareness of both the sectarian and ethnic violence that plagues the regions and the low media literacy of the Indian subcontinent, I saw the deepfake as something altogether more cynical—an attempt to deceive millions of voters with inauthentic content passed off as original ahead of an election. Admittedly, the content was relatively benign, but one could argue that by pretending to speak in the many different languages of the Indian subcontinent a politician is able to manipulate long-existing, and often deadly, ethnic and sectarian tensions. Whatever the motivation behind the video, it marked the first moment a deepfake blurred the line between reality and fiction in an organized election campaign. It will not be the last.

"THE LIAR'S DIVIDEND": A DEEPFAKE COUP IN GABON

We may be at the very start of the deepfake story but, even before we get to the bit where they become weapons of mass disinformation, the mere existence of deepfake technology is already being used to destabilize shared public discourse, and to undermine political opponents, truth and trust. One of the biggest threats already posed by deepfakes is the other side of the coin—giving everyone plausible deniability—the so-called "liar's dividend" that I introduced in Chapter 3—or by making the real seem faked.

In 2018–19, a "deepfake" was at the heart of a failed military coup in Central Africa. The people of Gabon had not seen their president, Ali Bongo, in public for months. As the heir to a dynasty that has ruled the oil-rich African nation for over

50 years, Bongo took over from his father in 2009. Gabon has been relatively stable in its recent history compared to other neighbouring countries, although Bongo's re-election in 2016 was marred by claims of fraud and violent protests. He had plenty of political rivals.

Fast-forward to October 2018 and Bongo was on a visit to Saudi Arabia to attend an investment summit. While he was there, the Saudi state agency reported that he had been hospitalized. At home in Gabon, there was little information about Bongo's condition. First, the administration kept schtum but then proceeded to give inconsistent reports about Bongo's health, first saying that he had suffered "severe fatigue," then updating that to "bleeding" in November. It released a video of Bongo, but without any sound. Intense speculation as to Bongo's whereabouts and health began, with suggestions that he had died, or that he was being replaced by a body double. In the face of the government's silence, these theories might have been encouraged by Bongo's political opponents. Finally, in early December, the Vice-President said that Bongo had suffered a stroke, and that he was in "bad shape."[13]

As the rumours reached fever pitch, the government attempted to instil calm by releasing a video of Bongo delivering his traditional New Year's address. It was his first public statement since rumours of his ill health began and was intended to prove that he was alive and well. But all was not well: Bongo looked strange in the video. It can still be viewed on YouTube:[14] his face barely moves, and it looks unnaturally smooth, with the wrinkles he used to have on his brow and between his eyes completely gone. His eyes look unusually wide, while the right eye looks bigger than the left. As Alexander W Dromerick, a neurologist, later told *The Washington Post*, Bongo's appearance was consistent with someone who had suffered a stroke or a brain injury. Dromerick also suggested that Bongo might have had cosmetic procedures, like botox, to try to alleviate some of the effects of the stroke.[15] If that were the case, it would help explain his wide-eyed, strange appearance.

The video did nothing to quell the rumours. Instead, it supercharged them. As Bongo's critics increasingly peddled theories that something fishy was going on, people started to believe that Bongo had been replaced by an actor in the video, or that the video was a deepfake. Bruno Ben Moubamba, a Gabonese politician who had run against Bongo in the previous two elections, was prominent in popularizing the deepfake theory. He said his view was driven by the fact that Bongo's face and eyes seemed "immobile" and "almost suspended above his jaw." He also rightly noted that Bongo's eyes moved "completely out of sync with the movements of his jaw."[16] When the deepfake theory was published in an article in the *Gabon Review* in January 2019, the confusion of the Infocalypse finally spilled over into real life.[17]

Gunshots rang out at the Gabonese national radio station at 3a.m. on 7 January 2019. At first, those living close by thought it was youths playing with fireworks. Soon, it became apparent it was the start of an attempted coup d'état, which was being broadcast live to the nation. You can find the footage online: a military commander sits in a chair in a radio station, dressed in a green beret and military fatigues. He is flanked by two guards, rigid and holding machine guns. He starts his address to the nation, "Gabonese, Gabonese. Dear compatriots, I am Lieutenant Kelly Ondo Obiang, deputy commander of the Republican Guard's honour campaign." He continues:

Once again, those fiercely holding onto power continue to use and to back President Ali Bongo Ondimba to keep in place an invalid who has lost his physical and mental faculties. The dreadful spectacle of this New Year's Speech is bringing shame in the eyes of the world to our country that has lost its dignity.[18]

The army alleged that Bongo had lost control and believed that the footage of the President's New Year address had been manipulated. Although the coup failed in less than 24 hours, it shows how quickly the information environment of the Infocalypse can spiral out of control, with potentially devastating consequences. In this case, the rumours concerning Bongo's ill health were being exploited by political rivals. But the same tactics could be used, of course, by state actors to intimidate and oppress political opposition or the general population.

Ali Bongo has since appeared in public walking with a cane. And while there has been no public confirmation of whether or not he suffered a stroke, it seems the most likely explanation. Forensics experts, including DeepTrace, the deepfake detection company based in Amsterdam, think that the New Year's address was an overly staged video of a sick man rather than a deepfake. But this incident highlights how vulnerable our information ecosystem is, not only to deepfakes but also to being corroded simply because we know deepfakes exist and there is confusion about the difference between truth and reality.

FAKE SEX SCANDAL?

For another example of how deepfakes can be used to make both the fake seem real and the real seem fake, in a political context, let us look to Malaysia, where another "deepfake" video went viral in the summer of 2019. In it was Mohamed Azmin Ali, the Malaysian Minister of Economic Affairs, having sex with Haziq Abdul Aziz, a male aide to a rival minister.

Days after the video went viral, Haziq released a Facebook statement confessing it was him and Azmin in the video. He alleged that they had been in a sexual relationship for over three years, accusing Azmin of recording the videos without his knowledge or consent. He wrote of Azmin:

> I know you are a sick man because only you could have recorded the videos for personal collection after inviting me to your hotel room on all occasions. I am worried more videos will be leaked, and the luxury you have as a minister is deniability, whereas my future is over. You even have the prime minister pledging his full support before any formal investigation has been done.[19]

Same-sex activity is illegal in Malaysia, with a colonial-era criminal ban on sodomy punishable by fines, caning and even prison sentences of up to 20 years. Sex scandals to take down politicians have been a hallmark of Malaysian politics since the 1990s. Politicians have been imprisoned on sodomy charges in the past, so this was a very serious affair. As the scandal broke, Haziq was arrested, but Ali and his supporters (including the Malaysian Prime Minister) argued that the sex tape was a deepfake launched to destroy his career. "This is 'cooked up,'" said the Prime Minister, "this was done by people who have political agendas."

As the nation debated whether or not the video was real, the video was combed over in excruciating detail for clues to its authentic or fake nature. One report in the Malaysian media, with the headline "Is it Azmin or a deepfake?" speculated that the video could be a deepfake because, "the Azmin character in the video was more interested in listening to a BBC News report about elite soldiers protecting French Guinea whose precious ecosystem is under threat from illegal gold mining than receiving pleasure." It added that the 'actors' movements were not voracious and wild."[20] Forensics experts, however, have not been able to find any indication that the video has been tampered with—let alone that it is a deepfake.

Nevertheless, Aziz is still a senior politician in Malaysia: he remained in the Cabinet and is now serving as the Senior Minister for Economy and the Minister of International Trade and Industry. If the video is authentic, then Aziz is enjoying the "liar's dividend." The bad actors of the Infocalypse will benefit the most from the blurring of reality, with the power to attack anyone, but deny everything. Not only will this further undermine trust, but it also gives those in positions of power a means to avoid accountability.

In the democratic West, the Infocalypse is well advanced, but defences still exist. As I discuss in Chapter 7, we are also starting to fight back. In the non-Western world, countries like India, Myanmar, Gabon and Malaysia, which arguably have fewer defences, bad and untrustworthy information may have even more serious and dramatic consequences. The point at which society can no longer cope may be closer. Ultimately, however, it impacts us all: democracy or not, the Infocalypse knows no bounds.

5

Deepfakes
in the Wild

The Infocalypse is already enabling scammers and criminals to thrive. Individuals and companies are vulnerable to these increased and evolving threats. While these sorts of attack have existed since time immemorial, the Infocalypse has made them easier to perpetrate, more potent and more widespread. Deepfakes will be these bad actors' next weapon.

The plot was bold and bizarre. In 2016, a gang of conmen stole over €50 million by posing as French Defence Minister Jean-Yves Le Drian. They relied on the power of audiovisual communication, contacting wealthy individuals through phone and video calls, and asking them to fund "secret" French government missions. As audacious as the plot was, the tools were relatively low-tech. One of the crooks put on a silicone mask of Le Drian's face and sat behind an official-looking desk. A French flag hung in the background. Then, he asked for millions of euros in donations. If you look up "Le Drian Plot" in a Google image search, you will get a visual of what the targets would have seen in the video call. It is not even close to the level of the deepfake examples we have seen in earlier chapters. The man wearing the silicone Le Drian mask looks like some kind of nightmarish vision, pallid-skinned and with bizarre black hollowed-out holes instead of eyes. Still, three people, all ostensibly savvy and successful business leaders in real life, ended up taking the bait. They included the Aga Khan, the spiritual leader of Ismaili Muslims, who made five transfers for a total of €20 million to accounts in Poland and China, and İnan Kıraç, a Turkish business magnate, who wired more than €47 million, allegedly to pay for the ransom of two journalists held hostage in Syria.

The success of what might seem like a faintly ridiculous scam is testament to the power of audiovisual communication. As discussed in Chapter 1, we are simply not conditioned to think of audio and video as media that can be subverted. Time and time again, we (even the richest, best resourced and best protected among us) fall for impersonation scams. In early 2020, Prince Harry was duped on a phone call by a pair of Russian pranksters, pretending to be the young environmental activist Greta Thunberg and her father. In a diplomatic faux pas, Harry divulged that he thought U.S. President Donald Trump had "blood on his hands." He was even prompted to discuss what was one of the hottest global news stories at the time, so-called "Megxit," or his departure with his wife Meghan from their official role as members of the British royal family.[1] If a dodgy puppet peering down a video link was good enough to convince some of the world's richest men to part with millions, and a pair of pranksters doing a terrible impersonation of Greta Thunberg could persuade Prince Harry to discuss deeply personal matters, then it is reasonable to conclude that we are not prepared for deepfakes. As discussed in Chapter 1, deepfakes extend beyond simple media manipulation. Because they can be generated from scratch on training data, they give criminals and fraudsters the ability to effectively steal and use our biometrics: they can take our image and our voice to do and say things that never happened.

Audio fraud is an extremely powerful tool used by scammers around the world, and AI is going to help them. AI is already very good at generating human voices in deepfake audio form. Take a look at the "Vocal Synthesis" (VS) account on YouTube. An anonymous YouTuber set up this VS account in August 2019 and has quickly amassed almost 7 million views. All VS does is generate synthetic audio clips of celebrities and politicians by using an open-source AI software developed by Google called Tacotron 2.[2] It is clear that VS wishes to entertain rather than harm, but because it is appropriating another person's voice—including bringing people back from the dead—the ethical and legal implications quickly get murky.

The most popular video clip on VS's YouTube channel resurrects the voice of President John F Kennedy from beyond the grave. When I first saw the title, "JFK reads the Navy Seals Copypasta," it made no sense to me. Copypasta? I clicked play. Astonishingly, I hear JFK's distinctive voice, "What the fuck did you just fucking say about me, you little bitch?" it starts. "I'll have you know I graduated top of my class in the Navy Seals, and I've been involved in numerous secret raids on Al-Qaeda, and I have over 300 confirmed kills." What? It sounds a bit tinny, or robot-like, but it is definitely JFK speaking. That is his accent, his cadence and his tone. Close to perfection.

"Navy Seal Copypasta" is a well-known Internet meme which makes fun of people posing as "tough guys" on the Internet. The meme originates from a viral 2012 post in which the author was doing exactly that—posing as a "tough guy." Responding to another Internet user's comment, he made a series of ridiculous claims and grandiose threats, including that he was a former Navy Seal with a long history of combat experiences and "300 kills." The post was full of comical typos and hyperboles such as "Gorilla Warfare" and "I can kill you in over 700 ways with just my bare hands."[3] Now, thanks to AI, this legendary Internet rant had been rendered in JFK's voice. The accent is perfect. "I am trained in gorilla

[sic] warfare and I'm the top sniper in the entire U.S. armed forces," he says. "You are nothing to me but just another target," he continues in JFK's distinctive Massachusetts accent ("just anothah tahgeht"). The JFK Navy Seal Copypasta rant runs for a full 1 minute and 44 seconds.[4]

Soon, AI will be able to overlay this voice with video, so that you'll not only be hearing JFK speak but also be able to see his mouth forming the words, his eyes blinking, his head moving and his limbs gesturing. Whereas the photographic manipulators working for Stalin were able to "unperson" just those Soviet politicians who had fallen out of favour, synthetic media can rewrite the historical record completely. Other popular VS audio clips include President George W Bush rapping the sexually explicit lyrics to 50 Cent's hit "In da Club" in his Texan drawl, including the line "I'm into having sex, I ain't into making love." Aside from JFK, VS also resurrects other dead American presidents, such as Franklin D Roosevelt and Ronald Reagan.[5]

In April 2020, the rap superstar Jay-Z brought copyright infringement claims against VS. The YouTuber had created a clip in which an AI-generated "Jay-Z" raps the "To be or not to be" soliloquy from Shakespeare's *Hamlet* and the Book of Genesis from the Bible. Jay-Z demanded it be taken down, as it was using his voice without his consent. VS sent a response via their YouTube channel, cloning the voices of Donald Trump and Barack Obama to say they were "disappointed that Jay-Z" decided to "bully a small YouTuber in this way."[6] This tussle between Jay-Z and VS is an early indicator of the type of challenges around privacy, security and consent that will become more commonplace as synthetic media become widespread in the Infocalypse.

Another early example is the case of Dr. Jordan Peterson. The famous public intellectual launched legal action against a website, NotJordanPeterson.com, that allowed users to generate deepfakes of his voice. He was especially alarmed by how it was used by his detractors to ridicule him and paint him in an unflattering light. One journalist used the website to generate deepfakes of Peterson reading parts of the "SCUM Manifesto,"[7] for instance. Written by the radical feminist Valeria Solanas in 1967, the SCUM Manifesto is deeply disturbing and violent. In it, Solanas argues that men are a "biological accident," that women must "immediately" begin to reproduce "without the aid of males" and that "deep down, every man knows he's a worthless piece of shit."[8] Peterson objected to his voice being used in this way, because of the violent ideology advocated by Solanas. A year after she wrote the manifesto, she shot and wounded the artist Andy Warhol. Though Warhol survived the attack, he had to wear a surgical corset for the rest of his life.[9] Peterson wrote about how disturbing he finds deepfakes in August 2019:

> I'm already in the position (as many of you soon will be as well) where anyone can produce a believable audio and perhaps video of me saying absolutely anything they want me to say. How can that possibly be fought? More to the point: how are we going to trust anything electronically mediated in the very near future (say, during the next presidential election)? [...] Wake up. The sanctity of your voice, and your image, is at serious risk. It's hard to imagine a more serious challenge to the sense of shared, reliable reality that keeps us linked together in relative peace. The deepfake artists need to be stopped, using whatever legal means are necessary, as soon as possible.[10]

EVERYONE IS AT RISK

Peterson is right. We need to wake up. It is unsurprising then that we are starting to see the deepfake scams of the future. In March 2019, *The Wall Street Journal* reported that a British energy company had been defrauded out of €250,000 through the use of deepfake audio. The company remains anonymous, but its insurers shared the events with *The Wall Street Journal*, claiming the scammers had used AI to mimic the voice of the company's German CEO.[11] Using this voice, they called a senior employee and directed that he immediately transfer €250,000 to the account of a purported energy supplier. (Independent experts have not verified the audio, so we cannot say with certainty whether or not it was a deepfake.) The employee thought it was an unusual request but complied because he believed he was speaking to his boss. It was only when he was asked to send another €250,000 that the alarm was sounded. By the time the banks and authorities were involved, the money had vanished and the trail was dead.

If they *had* used AI to clone the CEO's voice, we understand how they could technically do this. They would have had to gather personal data to train their AI algorithms. In this case, the training data would have been the voice of the German CEO. Given his prominent position, this might have been publicly available and easy to access. Perhaps he had made a speech which was on the company's website, YouTube or LinkedIn. Maybe he was featured in audio or video material on social media, such as an interview with a news channel. Perhaps he appeared somewhere on social media in his personal capacity. Even if the CEO had not posted anything himself, it might have been posted by someone else.

With the ability to steal someone's voice and likeness, the age-old impersonation scam has been turbo-charged. Within four months of the German CEO case hitting the headlines in March 2019, the cybersecurity firm Symantec reported that three other companies had fallen victim to similar ploys, with AI being used to clone voices and call senior financial officers requesting urgent money transfers. Symantec didn't reveal the names of the businesses but confirmed that millions of dollars had been lost. The true cost of fraud is difficult to quantify, but it is estimated to be in the trillions of dollars. An annual report published by accountancy firm Crowe Clark Whitehill and the University of Portsmouth's Centre for Counter Fraud Studies put the cost of global fraud in 2019 at \$5.127 trillion, with losses rising by 56 per cent in the last decade.[12] This jump coincides with the emergence of the Infocalypse. This trend will likely continue as deepfakes become more common. I spoke to Matthew F Ferraro, a former intelligence officer, who is now counsel at the international legal firm Wilmerhale, which warns about the extensive risks to business by deepfakes and disinformation. While "disinformation" currently falls under general risk mitigation, Matthew tells me that it is only matter of time before businesses have to invest in more specialized tools and strategies to actively combat these specific risks. Experian, the consumer credit agency, agrees. At the time of writing, it predicts that 2020 is the year in which deepfakes will become a more common disruption threat to large commercial enterprises. Even if it does not become widespread in 2020, it is surely only a matter of time. Of course, criminals are going to use deepfakes. They have the potential to lead to big-time pay-outs.

It is not only companies that need to be concerned. Individuals are going to be attacked, too. The AI powering synthetic media is advancing at such a rate that less and less training data is needed to generate deepfakes. This is true for both video and audio. One company launched in 2017, Lyrebird, claims to be able to generate realistic deepfake audio with just a few minutes of training data.[13] In future, that may become seconds. (Lyrebird has since been acquired by Descript, a company that is developing technology to edit audio in the same way text can be edited.) This means that it is not only high-profile individuals in the public eye (for whom training data can usually be easily accessed) who are in the firing line. We are all potential targets. If you are a prolific social-media user, your content is available for the taking. Even if you are not a social-media user, you may feature in content posted by your friends or family. You may have been filmed or photographed in a professional setting. Your phone might have been hacked, and private photos and videos could be stolen to generate a deepfake.

It is no exaggeration to say that if you have ever been recorded at any time in any form of audiovisual documentation, be that a photograph, a video or an audio recording, then you could theoretically be the victim of a deepfake fraud. These deepfakes will be used in a number of ways to defraud private citizens, from infiltrating our online banking to crooks posing as family members or a friend in distress. Elderly and vulnerable groups have traditionally been targets for individual fraud as they are supposedly easier to fool. However, with deepfakes, even the most discerning and savvy among us could be duped.

DESTROYING BUSINESSES AND LIVES

All organizations and individuals are more vulnerable to fraud in the new information ecosystem, and we are all now also potential targets of mis- and disinformation that could ruin our reputations, businesses and lives. Again, although reputational risk through targeted disinformation and misinformation are old threats, the potency of these threats has strengthened significantly in the Infocalypse. The cost of the Infocalypse to businesses is already huge. In 2019, the Israeli cyber-security firm CHEQ published a report with the University of Baltimore putting the cost of disinformation on the Internet against business at an annual $78 billion.[14]

In 2019, a deepfake was used by someone trying to infiltrate a set of high-profile investors. Tesla, the electric-car company, is one of the best-known brands in the world, thanks in part to the profile and infamy of its founder, Elon Musk. Just like Musk, Tesla is "Marmite"—"love it or hate it"— and investors have strong views on the company's prospects. On the one hand, the "bullish" investors think the company is undervalued, as it will change the future of transportation. On the other, the "bearish" investors think it is overvalued and facing inevitable collapse. The stakes are high in this market tug-of-war, with huge sums of invested money involved.

For now, the bullish investors still have the upper hand. Since the end of 2019, Tesla stock price has skyrocketed. By the end of January 2020, the S&P 500 (an index of America's 500 largest publicly traded companies) was up 3 per cent whereas Tesla stock was up 30 per cent. That same month, Tesla hit a market valuation of $100 billion.[15] Since the start of the year, the short sellers have lost billions. One figure put their January losses alone at $9 billion.[16] As the *Financial Times* reported, the losses Tesla inflicted on short sellers in January "were the biggest of any S&P 500 company [...] The losses were more than four times the $1.3bn drop in the value of short positions in Apple."[17]

As the battle between Tesla and its short sellers escalated, Musk made his contempt of the latter public. He often uses Twitter (where he has 36 million followers) to berate and goad them. With literally billions at stake, Tesla short sellers have justifiable cause to feel jittery about their position, not least because they could be dragged into a public war of words with Musk. In this fraught atmosphere, in March 2019, Maisy Kinsley, a senior Bloomberg journalist, connected with 195 people on LinkedIn, who were all investors tracking Tesla stock. She also began following prominent Tesla short sellers on Twitter (Figure 5.1 below). She messaged several of them, pressing them for personal and financial information. One

Figure 5.1 "Maisy Kinsley's" Twitter page

of the short sellers looked into Kinsley and grew suspicious. Although she had a professional-looking website and a LinkedIn profile, there were no stories filed under her name on the Bloomberg website or on any other credible news site. It was a bit strange for a journalist to have no public writing history. He raised the alarm, warning jittery short sellers that Maisy was a cover for someone potentially seeking to harvest market-moving information. It turned out that Maisy wasn't a journalist at all, but someone posing behind a fake persona with a GAN-generated deepfake image. This could have been downloaded for free from www.thispersondoesnotexist. com. Bloomberg later confirmed that they had no employee named Maisy Kinsley.

Even though "Maisy" did limited damage, this is a warning shot. As the AI becomes more advanced, markets and investors are vulnerable. Deepfakes could be used to undermine the reputations of individual investors, for example through a "leaked" audio tape where an investor "admits" to wrongdoing. As the case of Maisy reveals, deepfakes could also be used to extract sensitive investment information. The Maisy of 2019 used one deepfake image to build a website, a LinkedIn profile and a Twitter profile, but in future Maisy could build credibility as a fake persona by populating her website and social media with deepfake video of her interviewing business and investment leaders, for example, helping her to gain the trust of targeted individuals in order to "catfish" them. It will be possible to generate an entirely fictional history for an AI character like Maisy as she interacts with real people.

Tesla is also an interesting case study owing to the antics of Musk. With such an erratic CEO, it is possible to see how a deepfake could be deployed to move markets, for example through a "leaked" video in which "Musk" reveals financial information. His behaviour is already responsible for stock price fluctuations. In August 2018, he infamously tweeted that he was "considering taking Tesla private" at $420 a share, saying he had "funding secured" to do so. It turned out to be false information, and *The Wall Street Journal* later reported that the 420 reference might have been a joke for Elon's girlfriend (as "420" refers to smoking cannabis). At the time, the tweet was material information to Tesla investors and caused a market rally around Tesla's share price. The Securities and Exchange Commission (SEC) found Musk guilty of misleading investors in a market-moving event, subsequently fining him and his company $20 million each. In a deal with the SEC, Musk agreed to step down as Tesla Chairman and have his tweets on Tesla vetted before posting them. That clearly did not happen. On 1 May 2020, Musk tweeted, "Tesla stock too high imo," wiping $14 billion off Tesla's valuation.

One securities analyst later told Reuters that this was a case of "Elon being Elon," that it was a "headache for investors" and that Wall Street was "clearly frustrated" by the "hot button issue" of his tweeting.[18]

Given Musk's erratic public history around his company and its valuation, it is easy to see how deepfake content targeting Musk could snowball into a market-moving event. In future, there will be many more attacks on private companies, be they as a result of a conspiracy going viral or of an organized campaign to destroy a company's share price or ruin a CEO's reputation. This is systematically being turned into a commercial practice, with "black PR firms" now offering "disinformation for hire" services to attack the credibility of enemies and competitors. A BuzzFeed News investigation found that the disinformation tactics commonly associated with state actors, such as Russia, are increasingly being utilized against private-sector firms. In its investigation, BuzzFeed found one black PR firm that promised "to use every tool and take every advantage available in order to change reality according to our client's wishes."[19] It seems inevitable that deepfakes will be used to attack companies and distort markets when they are widely available for criminal use.

ALL WOMEN

When it comes to attacking individuals, the first and most malicious use of deepfakes is already doing that. It is not only political journalists like Rana Ayub or famous actresses like Scarlett Johansson who are in the firing line (see Chapters 1 and 4). Deepfake porn is being directed against all women: our wives, daughters, sisters and mothers. Deepfakes are the latest evolution of non-consensual and fake porn involving normal women. Even before AI, the phenomenon of fake porn destroyed lives. Whatever its guise, the exposure, humiliation and fear that come along with being targeted in this way are devastating its victims. It makes it difficult for them to go online, get or keep a job, and feel safe. It is an invasion of their most intimate lives and right to security. Danielle Citron, a leading privacy scholar and professor at Boston University law school who has spoken both about deepfakes generally as well as their use in non-consensual porn, believes that deepfake porn is a harbinger of what is going to emerge as a much broader civil-rights problem, as our likeness and voices are stolen and our rights to privacy and security consequently undermined.

The experience of one young woman from Australia, Noelle Martin, provides some insight into the devastation that non-consensual fake porn can wreak. Her ordeal started when she was a teenager, just 17 years old. One day, as almost all of us have done, she decided to google herself, using an image which she had posted on her social media. What she saw made her heart stop. She later described that awful moment:

> In a split second, my screen was flooded with that image and dozens more images of me that had been stolen from my social media, on links connected to porn sites. On these sites, nameless, faceless sexual predators had published highly explicit sexual commentary about me and what they'd like to do to me. "Cover her face and we'd fuck her body," one person wrote. They also published identifying information about me: where I lived, what I studied, who I was.[20]

Noelle was targeted before deepfake existed, so the tools were more rudimentary, such as photoshopping her face onto the bodies of porn stars having sex. Noelle's story is not unusual. She is one of thousands of ordinary women who are being exploited online. Even worse, Noelle found that no one could help her. When she went to the police she was told there was not much they could do. The sites were often hosted abroad, and the perpetrators were anonymous. Desperately, she contacted the sites one by one, notifying the webmasters to get everything deleted. But Noelle had slipped into a fake-porn whack-a-mole dystopia. She managed to get some content deleted, but new porn kept popping up. One webmaster blackmailed her, telling her she would only get the fake porn removed, if she sent him nude photos within 24 hours. After years of battling the online bullies, Noelle decided to speak publicly about her experience in 2016. Her story became key to new legislation introduced in Australia in 2019 which criminalized the non-consensual distribution of intimate images.

Unfortunately, the problem is moving faster than the solution. The deepfake porn ecosystem is already being commodified with an array of deepfake creation services and apps. For example, the DeepNude app was launched in summer 2018—one could download it for free and upload a picture of a woman fully clothed and it would then generate a fake image of that woman "stripped" or naked. Using a GAN, the app took 30 seconds to generate the naked image. No skill was required on the part of the user, apart from being able to upload a photograph. The images the app produced were obscured by a large watermark, and users had to pay $50 for the licensed version of the app to remove this. The licensed version still contained a small watermark denoting the image as "fake." The DeepNude app was only trained on a data set of naked women, so it did not work for men. If you tried to upload an image of a man, it would generate a vagina over his pants rather than a penis.

The app was discovered by the journalist Samantha Cole from Motherboard. She slammed it in a piece entitled "This Horrifying App Undresses a Photo of Any Woman With a Single Click."[21] Unfortunately, Sam's piece alerted the wrong people to the problem. As soon as it was published, interest spiked. Under a stampede of downloads, the DeepNude server crashed, causing it to go offline. The developers estimated that they had almost 100,000 users in the 24 hours after the Cole piece went online. Eventually, DeepNude came under so much pressure that the developers took the website offline, declaring "the world is not yet ready for DeepNude app." The developers realized, however, that they had hit a jackpot. The following month, they put DeepNude up for sale, where it went for $30,000 to an anonymous buyer in an online auction. The software behind the app is now being repurposed and distributed. Henry Adjer, the head of Threat Intelligence at DeepTrace, told me how the software is now available for download on various torrenting websites. DeepTrace also identified "two new service portals that offered allegedly improved versions of DeepNude," with charges ranging from $1 per photo to $20 for a month's unlimited access.

The moment DeepNude was made available to download "it was out of the creators' control, and now is highly difficult to remove from circulation." The software will likely continue to "spread and mutate like a virus, making a popular tool for creating non-consensual deepfake pornography of women easily accessible and difficult to counter."[22] The DeepNude app used GAN technology to generate images of naked women. The face-swapping techniques I described in Chapter 1 are used to create fake porn video and still require some level of skill to do well. However, as synthetic-media generation develops, there will be new and better ways to create deepfakes that will make the output of apps like DeepNude and face-swapping look rudimentary.

The founder of the World Wide Web, Sir Tim Berners-Lee, has noted with concern that women and girls face a "growing crisis" of online harm, with sexual harassment, threatening messages and discrimination making the web an unsafe place for them to be. In his own words, "The web is not working for women and girls."[23] But the reality is that, in the Infocalypse, it is not only women and girls who will face a growing crisis of online harms but also men and boys. While the deepfake porn phenomenon has so far been almost exclusively targeted at women, homosexual deepfake porn could, for example, cost someone their life or liberty in certain parts of the world. It won't be long before it is used in this way.

CHAOS AND THE CASE OF PIZZAGATE

While the conditions of the Infocalypse increase the vulnerability of all organizations and individuals to deliberate and targeted attacks, the chaos of the Infocalypse also makes us more vulnerable to becoming collateral damage. The snowball effect of false information going viral, even without malicious intent, has the power to destroy lives and businesses. One infamous pre-deepfake incident, now known as "Pizzagate," is a cautionary tale for how things can escalate in our dangerously weaponized information environment

The Pizzagate story starts with the Russian email hack of the Hillary Clinton presidential campaign in 2016, and it very nearly ended in disaster. A month ahead of the election, John Podesta, chair of Hillary Clinton's presidential campaign, had a stolen cache of his emails dumped in full onto the "hacktivist" WikiLeaks website (where they can still be downloaded today[24]). Podesta was a diehard Democrat campaigner who had worked for both the Clintons before falling prey to "Guccifer 2.0," a Russian military-intelligence persona who claimed credit for the hack. These emails documented everything from the banal—such as a risotto recipe—to the politically explosive, such as transcripts of the private speeches Hillary Clinton had delivered on Wall Street for megabucks.

It was not long before the Internet was poring over every single detail, and a very strange conspiracy theory started to emerge on places like Reddit and 4Chan. Apparently, the emails provided evidence that John Podesta, Hillary Clinton and Barack Obama were members of an elite cabal of paedophiles, operating out of the basement of a pizza parlour in Washington, DC. How had this conclusion been reached? Well, the emails made mention of social gatherings involving "pizza," and "cheese pizza" is apparently code for child pornography on the dark web.

The conspiracy theorists saw patterns everywhere. For example, the pizza parlour was called Comet Ping Pong, and the initials "CP" (as with "cheese pizza") stood for child porn. The place was owned by a man named James Alefantis, who was seemingly well connected to the liberal DC establishment—further "proof." When spoken aloud, "James Alefantis" sounds like *j'aime les enfants*, French for "I love children"—another "sign." The contents of Alefantis's

Instagram account stoked the flames of conspiracy still further, as it included photographs of children, one of whom was eating a slice of pizza. In another photo, President Obama was pictured playing ping-pong at the White House with a young boy. These pieces of "evidence" were interpreted as signs of Alefantis's sick and deranged sex ring.

Websites, blogs and forums dedicated to the Pizzagate conspiracy popped up. One anonymous WordPress website called "dcpizzagate" feverishly detailed why the theory must be true. In a long, rambling blog post, the website says of the emails, "If this is not speaking in code, then it makes absolutely no sense. Not a single person from Podesta's legal team nor mainstream media has been able to offer a coherent explanation on what this could mean if [it was] ACTUALLY pizza [...] related."[25] It then goes to describe how deranged it finds Alefantis's "PUBLIC Instagram" profile. "Let's take a look at some of his Instagram posts and please tell me it does not make your skin crawl,"[26] says the blog. It continues by giving a blow-by-blow interpretation of images taken from Alefantis's Instagram account to weave a narrative of why he must be operating a paedophile ring. And it is not just a paedophile ring, according to this blog post, but one that is steeped in satanic occult ritual.

It all sounds crazy—and, of course, it is. It is all completely whacko. But somehow Pizzagate kept getting bigger. Reddit had a community of 22,000 users dedicated to the bogus theory. Small numbers of protestors started gathering outside Comet Ping Pong, harassing Alefantis and others accused of being part of the sex ring. Alefantis later described the abuse:

Messages and comments poured in. At one point, I was getting about 75 personal messages on social media a day. My response was to shut everything down—delete the comments, try not to respond. I thought, obviously this has to end eventually. But it only kept escalating. Many of the messages were violent—"I have a gun. I want you to die"— and gory—"I pray that someone comes in with an assault rifle and kills everyone inside Comet. I want to cut open your guts and watch them spill out on the floor." I'd just shut my laptop and go on with my day. It got scary when users started performing what I'll call "citizen's investigations." Users sifted through my social media profiles and started messaging everyone who'd ever liked or commented on a post of mine. I started getting calls from friends, family and customers who said they were getting harassed online as well.[27]

Then it escalated to the next level. The Podesta emails were published on WikiLeaks in October and November 2016. Alefantis thought the abuse would stop after Donald Trump was elected, but it did not. On 4 December 2016, a gunman walked into Comet Ping Pong, armed with an AR-15 rifle, and fired several shots. Thankfully, no one was hurt. When he was satisfied that there were no children (indeed, there was no basement!), the shooter, Edgar Maddison Welch, left the premises and surrendered himself to the authorities. He told them he had driven to DC from his home in North Carolina to "self-investigate" the rumours of the child-sex ring. Welch has since been sentenced to four years in prison, but the online world of Pizzagate lives on: the conspiracy theorists saw Welch's attack as a decoy meant to divert them from the truth!

This is an astonishing example of how the Infocalypse facilitates the escalation and spread of this kind of nonsense. If a couple of leaked emails mentioning "pizza" can lead a gunman to storm a building looking to free children ensnared in an imagined child-sex ring, imagine what deepfakes could do. Imagine there had been a deepfake video of Clinton and Alefantis discussing the sadistic abuse of minors. There could have been an even more dangerous situation than one lone gunman. Things never went back to "normal" for Alefantis.

Some people have continued to believe the conspiracy. He is still being abused. As he later said, "From that point on, I was fearful. I was still receiving death threats. I started wearing a hat and sunglasses to leave the house. And the security people were like, 'Yeah. You're in danger.'"[28]

Pizzagate is an extraordinarily bizarre story, but the storm in which Alefantis found himself is another example of how the Infocalypse is shaping our reality. As our information environment becomes even more polluted and corroded, people and businesses are becoming more vulnerable, whether as potential targets or collateral damage.

6

Covid–19

a global virus

I could not have predicted that, as I was writing this book, the world was to be plunged into an unprecedented crisis that would perfectly exemplify all the trends of the Infocalypse. The coronavirus pandemic is sweeping the world. Those infected and dying because of this horrific virus literally cannot breathe. For the rest of us, it feels like an invisible executioner is tightening his noose around our necks, starving all of us of our lives and our livelihoods. This has never happened before and we do not know how it will end. It has forced us to confront our own mortality just as it shuts down our livelihoods. It is also a perfect case study of the inner workings of our increasingly dangerous and untrustworthy information environment.

THE ROLE OF CHINA

I first heard of Wuhan on a bitterly cold winter evening in late January 2020. It was the day the virus arrived in Europe. I was on my way to Sky News in London for the evening paper review and reading early editions of the next morning's headlines. A deadly virus causing severe respiratory illness had emerged in Wuhan. Within weeks, dozens of people had died. Chinese New Year, one of the biggest celebrations in the lunar calendar, was days away. Millions of people were due to crisscross the nation (and the world) to celebrate with their loved ones. The Chinese government decided to take radical action. Overnight, Wuhan and the surrounding regions were shut down. Some 25 million people were trapped inside the largest cordon sanitaire the world had ever known. But even Beijing yanking on the emergency brake could not prevent the novel coronavirus developing into a full-grown pandemic. In the months I have been writing this book, I have also noticed how the Covid-19 crisis serves to illustrate the key aspects of the Infocalypse.

In Chapter 2, I introduced the idea of the Infocalypse through a geopolitical lens. I described how one state actor, Russia, channelled the spirit of the Infocalypse long before it came into being. Now that we all exist in the Infocalypse, Russia has been thriving. It exploits the conditions of the Infocalypse to help it achieve its political objectives. True to form, Russia is using Covid-19 to divide and distract targets across the world, like a turbo-charged version of Operation Infektion. Almost as soon as the outbreak began, the Kremlin linked it to the long-running trope of the U.S. biological warfare programme. These narratives hearken back to the Cold War, when the Soviets accused the United States of co-opting "fascist technology" from the Nazis to genetically engineer diseases. The Kremlin has alleged that the virus is an American "bioweapon" to target the Chinese. By the time Covid-19 hit the United States, the message had morphed: the virus was now said to have been manufactured for monetary gain by U.S. pharmaceutical companies, which had supposedly already developed a vaccine.

The virus has also enabled the Kremlin to exploit the increasingly bitter relationship between Beijing and Washington. In line with its "divide and distract" strategy, Russia has promoted the narrative that Covid-19 is a Chinese-made bioweapon and that it is linked to the construction of Chinese 5G networks. In Ukraine, the Kremlin disseminated stories to stoke fear that the situation was worse than it actually was. This led to a violent riot in a village in rural Ukraine when Ukrainian citizens who had been repatriated from Wuhan arrived home. Protestors blocked roads and threw stones at the buses carrying the evacuees. The national guard had to be brought in to calm the situation.[1]

Russian information operations went into overdrive the day Boris Johnson, the British Prime Minister, was admitted to hospital as coronavirus ravaged his body. False stories circulated that Johnson was on a ventilator and could not breathe unassisted, provoking a furious denial by the British government.

However, the Infocalypse is no longer the exclusive domain of Russia. Other state actors are increasingly seeking to exploit it, too. Covid-19 marks a turning point for China's tactics. Previously, Chinese information operations tended to be about extreme control through censorship and narrative-creation. However, from March 2020 China has also started to embrace a strategy that has more of a Russian flavour, seeking to sow confusion around the origin of the virus. Disinformation analysts have traditionally differentiated between Russia's "chaos" and China's "control" strategies, so it is a significant development that China is now mixing up its approach.

China's information operations in response to Covid-19 are comprised of three broad strands. First, there is censorship. In the earliest stages in Wuhan, the Chinese authorities censored news about the emerging pandemic and any criticism of the central government. They did this by downplaying the risk, silencing early whistle-blowers, and underreporting the number of deaths. Dr. Li Wenliang, an ophthalmologist working in Wuhan Central Hospital, became a tragic figure in this story. When he first tried to raise the alarm in late December 2019, he was hauled before the authorities and made to sign a statement that he had made a false statement that "disturbed the public order."[2] In a bitter twist of fate, Dr. Li subsequently died of Covid-19, leaving behind a young son and pregnant wife. Before his death, he gained worldwide recognition as a hero. In an interview with *The New York Times* he said, "If the officials had disclosed information about the epidemic earlier I think it would have been a lot better."[3]

As Covid-19 spread like wildfire beyond China, the Chinese Communist Party (CCP) continued to suppress information, including on social media. One report from the University of Toronto suggests that keywords relating to the outbreak were censored on Chinese social-media apps from December 2019. By February 2020, all critical, and even neutral, information relating to the government's response was being blocked. Meanwhile, the Chinese Internet governance agency issued a public warning that anyone publishing "harmful" content and "spreading fear" on websites, platforms and accounts would be punished.

These actions put not only Chinese citizens at risk, but everyone. In a damning report, the Foreign Affairs Committee in the British Parliament said that the Chinese authorities had "obfuscated the data" and "deliberately misled" public-health bodies and other governments, thereby "obscuring analysis" that would have been vital in "the critical early stages of the pandemic."[4] In the words of the committee chair, Tom Tugendhat, rather than helping other countries prepare a swift and strong response, it was "increasingly apparent that [China] manipulated vital information about the virus in order to protect the regime's image."[5]

The CCP understands how much it stands to lose in the face of this epidemic. That is why it is stepping up its second strand of information operations: a concerted attempt to portray China as a responsible and benign actor. This has led to overinflated reports on how well the crisis has been controlled at home, and so-called "face-mask diplomacy," by which Beijing is sending aid missions, encompassing both equipment and personnel, to help other governments deal with the virus, particularly Western governments. I saw this clearly when I came across the photographs posted on social media by Xinhua, the state-sponsored media agency, showing boxes of Chinese aid arriving at Heathrow Airport in the UK. They were labelled "Keep Calm and Cure Coronavirus."[6] Similarly, when China is accused of obscuring the early data (which it did), Chinese ambassadors have responded by suggesting that a more effective way to behave would be "with great solidarity" rather than "finger-pointing."[7]

Thirdly, Covid-19 has also seen China seek to sow confusion, in particular around the origin of the virus. This started in March 2020, when Zhao Lijian, the Foreign Ministry spokesman, claimed there were "no supporting facts or evidence" for calling coronavirus a "China virus." He suggested that anyone doing so had "ulterior motives" to "make China take the blame."

On 12 March, Lijian posted a link on Twitter to what he described as a "very much important" article that falsely described the origins of the coronavirus as American. Lijian went on to accuse U.S. officials of not coming clean about "what they know" about the virus.[8]

This was followed by coordinated action by Chinese ambassadors and state-sponsored media outlets sharing theories that Covid-19 originated outside of China, or might have been brought into Wuhan by the U.S. army.[9] This approach can be seen as part of a new hawkish strategy from Beijing, intended to deflect attention from its own missteps.

Other state actors are also launching their own information operations around Covid-19, including Iran and North Korea. Although their capabilities and tactics differ, they are generally using Covid-19 to censor information at home, promote narratives denigrating the West, hack personal data and increase surveillance. The Iranian regime has been sending out desperately needed health information that is laced with spyware to track citizens' movements. It does not bode well that an increasing number of state actors are stepping up activity in the Infocalypse. This should serve as a warning shot about how these trends will develop and accelerate.

COVID-19, TRUMP AND THE 2020 ELECTION

In the midst of the pandemic, the dynamics of the Infocalypse are becoming ever more apparent in the West, too. Again, we can use the United States as a case study. Objectively Trump has handled the situation terribly, and is fighting the pandemic with his own brand of Covid-19 disinformation,

Trump's administration was caught unaware by Covid-19. Despite early-warning signs, it refused to prepare like other nations did. At the time of writing, the virus has cost over 100,000 American citizens their lives, with the highest infection rate in the world. The roots of this botched response began in April 2018, when the Trump administration dismantled the team in charge of pandemic response at the White House National Security Council (NSC). It also made repeated cuts to public-health agencies' budgets, including the scrapping of a programme called "Predict," a $200-million early-warning programme designed to alert it to potential pandemics.[10]

While countries such as Germany, New Zealand and South Korea took the impending pandemic seriously, preparing ICU beds, ventilators and plans to roll out testing and contact-tracing capabilities, Trump's government did not. The President was first questioned about whether or not he was concerned about a pandemic at the World Economic Forum in Davos, Switzerland, on 22 January. This was hours before Wuhan was put into lockdown and the first known cases emerged in Europe. "Not at all," responded Trump. "We have it completely under control."

Increasingly concerned by this cavalier response, the experts whom Trump had fired in 2018 wrote an op-ed in *The Washington Post*. They begged the government to prepare, to "stop a coronavirus outbreak before it starts," warning that "a pandemic seems inevitable." They pleaded that all patients with "unexplained pneumonia" be tested even if they had not travelled to China.[11] Still, Trump continued to ignore the brewing storm. Not only did he ignore it, he actively spread false information about Covid-19. At rallies in February, Trump made claims that the virus would "die in hotter weather" or that it would "disappear, one day [...]

like a miracle" and that "you'll be fine." He even downplayed the urgent risk by accusing the Democrats of politicizing Covid-19 as a new "hoax." His exact words were: "It's all turning, [the Democrats] lost, it's all turning. Think of it. Think of it. And this is their new hoax."[12]

By mid-March, the confirmed number of Covid-19 cases in the United States had doubled to 10,000. Still, Trump was claiming everything was fine. "We're doing great, our country is doing so great," he said. When asked if he took any responsibility for his administration's patently inadequate response, Trump responded, "No, I don't accept responsibility at all." By now the markets were in freefall, collapsing faster than at any time since the great recession of 2008. On 20 March, Goldman Sachs warned that U.S. GDP would shrink by 29 per cent by the end of the second quarter of 2020, and that unemployment would skyrocket to at least 9 per cent. Finally, the President realized he needed to engage. Three days earlier, on 17 March 2020, Donald Trump had told gathered reporters, "This is a pandemic." He then added, "I felt it was a pandemic long before it was called a pandemic."[13]

As soon as the severity of the crisis finally dawned on him, Trump went into disinformation overdrive. Months ahead of the election, his entire focus switched to making sure to shift the blame for the botched response over which he presided.

The problem for Trump is that Covid-19 reveals the dangers of Infocalypse (and his role in it) perfectly. Bad information is dangerous, period. But in a pandemic it literally costs lives. Because accurate information is needed to fight Covid-19, Trump has ironically found himself in a situation where he is bound to seek out public-health experts in order to formulate and instil faith in the government response. At the same time he actively continues to spread mis- and disinformation. The situation would be comic if the stakes were not so high.

On the one hand there is Trump, and on the other his scientific advisors. Dr. Anthony Fauci, the 78-year-old director of the National Institute of Allergy and Infectious Diseases, is the perfect counterweight to Trump. A member of the White House Covid-19 Task Force, he has been briefing the nation alongside the President. A two-second grimace and face-palm in reaction to Donald Trump's reference to the State Department as the "Deep State Department" during a briefing in mid-March made Dr. Fauci an Internet legend. Since then, he has been using his rising popularity to inform Americans about Covid-19, even as it makes his status within the administration more precarious. While Trump embodies the Infocalypse—a source of misleading and dangerous information—Dr. Fauci is the opposite—a source of reliable, factual information. This is the surreal reality of the Infocalypse in the United States, months ahead of the election: the leading public-health experts share a platform with the President to communicate to the nation, and they end up briefing against each other. There is no better symbol of this than when Trump stood at the White House podium suggesting that "disinfectant" could be delivered by an "injection inside [...] or, almost a cleaning"[14] as a cure for the virus while one of his long-suffering scientific advisors (in this case Dr. Deborah Birx) looked on in stony-faced silence.

Aside from making such dangerous claims about how to treat the virus, Trump and his allies have also sought to deflect blame for the pandemic at home by attacking the Democrats for distracting them with the impeachment process. Externally, the administration has launched aggressive information operations against China. This war of words between Beijing and Washington will have significant geopolitical implications for the whole world. On 24 January 2020 in Davos, two days after Trump had said he was not worried about a pandemic, he was upbeat about China. He tweeted, "China has been working very hard to contain the Coronavirus. The United States greatly appreciates their efforts and transparency. It will all work out well. In particular, on behalf of the American People, I want to thank President Xi!" In the background, Trump's aides were asking the President to do more to press President Xi on the issue of transparency regarding the virus, but Trump refused—twice.[15] At the time, he was trying to smooth over a long-running trade dispute between Beijing and Washington, and he was busy making complimentary public statements about how the relationship with China was "the best it's ever been" and how he has a "great relationship" with President Xi.

In reality, the already fraught U.S.–China relations were about to nosedive. This happened in tandem with the worsening health crisis. By March, Trump had changed his tune on the "great relationship" by calling Covid-19 the "Wuhan flu" and "China virus." While it might have been tactless diplomacy, objectively it is true: Covid-19 originated in China. Of course, this did not help calm the situation. Moreover, it was only the start of Trump's blame game. Since then, the accusations and tensions between Beijing and Washington have reached fever pitch, including an assertion by Trump that China's handling of the pandemic is proof that Beijing "will do anything they can" to make him lose his re-election bid in November.[16] Contradicting his own earlier statements, Trump is now accusing China of a cover-up, promising to hold China "accountable" and threatening a "massive investigation." In April, *The New York Times* reported that senior Trump administration officials were applying pressure on U.S. intelligence agencies to link Covid-19 to Wuhan labs. Fearful that the White House wanted to use this as a "political weapon" in an intensifying battle with China, they refused. While it is true that China suppressed information about the break-out, there is no evidence that it was created in a lab in Wuhan.[17] Instead, the U.S. intelligence agencies issued a joint statement that the virus was "not manmade or genetically modified." Hours later, Trump claimed to have seen evidence to support the theory that Covid-19 was made in a lab in Wuhan. Trump's message appears to have been effective. A YouGov and Yahoo News found that 58 per cent of Trump voters agreed with the statement that "Chinese scientists engineered coronavirus in a lab, from which it accidentally escaped."[18]

COVID-19 AND THE CONSPIRACY THEORISTS

Trump's disinformation around Covid-19 has been dangerous at home and abroad. Although he is a particularly important "influencer," he is not the only person spreading mis- and disinformation about Covid-19. The pandemic has driven conspiracy theorists into overdrive. Daniel Jolley, a senior lecturer in psychology at Northumbria University, explains that conspiracy theories are rooted in the belief that "a powerful group of people are plotting something for their own gain." Conspiracies are especially potent and persuasive in moments of crisis because they help people navigate them by attributing a reason to the crisis. As Jolley explains, people reach to conspiracy when "we feel the need to feel certain."[19]

In the U.S., QAnon, a group of conspiracy theorists who think there is a "deep state" conspiracy involving a cabal of elite paedophiles, are propagating the conspiracy theory that Bill Gates created Covid-19. (As we saw from Pizzagate in Chapter 5, there are a lot of online conspiracies about elite paedophiles who rule the world.) One version of this conspiracy claims that Bill Gates is plotting to use a Covid-19 vaccination campaign as a pretext to implant microchips into billions of people to monitor their movements. The Yahoo News/YouGov poll I referenced earlier found that 44 per cent of Republican voters believe this conspiracy, and that only 50 per cent of Americans say they would get a Covid-19 vaccine if one was developed.[20]

It should come as no surprise then that the Bill Gates theories have also found traction in the international anti-vax community, a growing group of people who choose to refuse vaccines against contagious diseases for themselves and their children. They do this for various reasons, including the belief that vaccinations cause autism. While vaccine hesitancy has a long history, it has gathered speed and popularity in tandem with the emergence of the Infocalypse, and we are now seeing the resurgence of emergency-level breakouts of diseases such as measles in the Western world. The anti-vax community is an easy target for hostile foreign states. Indeed, Russia exploits them just as it exploits African-Americans.[21] Even Covid-19 has not shaken the confidence of some anti-vaxxers. One influencer, the singer M.I.A, claimed that she would "choose death" over a vaccine.[22] The world's number-one tennis player, Novak Djokovic, suggested that he may not return to the sport because he "would not want to be forced by someone to take a vaccine" to travel.[23]

Meanwhile in the UK, the infamous conspiracy theorist David Icke has been expounding the theory that Covid-19 is being spread by 5G networks. For context, this is the same man who claims that the world has been hijacked by an interdimensional species of reptilian beings known as Archons, who pose as the world's elite, including the British royal family. The 5G conspiracies, which play on legitimate fears about national security and China's role in providing such critical information infrastructure, pre-date the current crisis. However, they have now evolved to encompass Covid-19. In the UK, this is taking a dangerous turn. At the time of writing, there have been over 70 arson attacks on network masts and over 180 reported incidents of abuse of key workers implementing 5G projects.[24]

Covid-19 has also seen its first deepfake. Arguably, it is the single best example of a politically damaging deepfake that exists "in the wild" to date. It was made by the Belgian branch of the environmental-activist group Extinction Rebellion (XR). It features a fictional speech from Belgian Prime Minister Sophie Wilmès, in which she claims that global epidemics like SARS, Ebola and Covid-19 are directly linked to the "exploitation and destruction by humans of our natural environment."[25] She appears to sing off the XR hymn sheet when she says:

> Coronavirus is an alarm bell we cannot ignore [...] Pandemics are one of the consequences of a deeper ecological crisis. We have failed as policymakers to grasp the seriousness of the ecological collapse. But today, the coronavirus crisis is making us aware of the depth of change required of us: we must change our way of life, and we must change it now.[26]

The XR activists responsible have not disclosed how they made the video, and although on Facebook they described the video as "fake," this was not clearly marked in the title. As the comments under the video show, many viewers thought it was authentic.

AUTOCRATS AND CRIMINALS

Dangerous and untrustworthy information surrounding Covid-19 is a global issue. In India, politicians of the ruling Hindu nationalist party, the BJP, are using the crisis to push their own nationalist pseudo-science, claiming, for example, that urine and dung from the sacred cow are remedies. In Brazil, President Jair Bolsonaro insisted on resisting lockdown and social-distancing measures, dismissing Covid-19 as a "little flu" and accusing the media of "hysteria." Coronavirus reached South America later than other continents, but at the time of writing Brazil has the second-highest infection rate in the world after the United States, with an exponentially rising death toll (the sixth highest in the world at the time of writing.[27]) A study by Imperial College London analysing the active transmission rate of Covid-19 in 48 countries showed that Brazil is the country with the highest rate of transmission ($R0 = 2\cdot81$).[28] Bolsonaro continues to foment confusion by openly flouting and discouraging the sensible measures of physical distancing and lockdown brought in by state governors and city mayors. When asked by journalists about the rapidly increasing numbers of Covid-19 cases, he responded, "So what? What do you want me to do?"[29]

Russia is facing a similar crisis. It briefly had the second-highest infection rate before being overtaken by Brazil. In March 2020, when it first appeared that Russia had not been as hard hit, Moscow rushed to send foreign aid abroad, including to the United States. Putin's government failed to use the lag time to prepare adequately. Within a few weeks, it was Russia that was importing ventilators from the United States.[30] Ironically, Putin displayed the same Covid-19 ineptitude that he was accusing Western leaders of.

But it is not just politicians, conspiracy theorists and state actors at play. The whole gamut of racketeers, thieves and cyber criminals has been animated by Covid-19. There are crooks, for example, selling fake remedies to profit from the pandemic. In March, Interpol, the international policing agency, used Operation Pangea, its programme to fight counterfeit and illicit health products, to intercept more than 34,000 fake coronavirus products. This included everything from counterfeit facemasks and substandard hand sanitizers to unauthorized antiviral medication.[31]

Cyber criminals are having a field day, too. These conditions are perfect for them to exploit individuals, organizations and businesses. With hundreds of millions of people in lockdown and desperate for information, it is like shooting fish in a barrel. Covid-19 has seen the exponential rise of phishing scams, where criminals harvest private data. Many of them are sending out emails pretending to be public-health authorities (such as the World Health Organization), to entice users to download malware that steals personal information and data like credit card numbers. As in the case of Iran, these are also tricks being used by state actors to install spyware. With people working from home, companies have had to set up their remote systems quickly, often compromising on security. This, too, has made it easier for cyber criminals to exploit Covid-19.

The dangers of our corroding information ecosystem are highlighted by the emergence of Covid-19. As there is so much unknown about this virus, the blanks can be filled in with bad and untrustworthy information. The Covid-19 crisis shows us the Infocalypse at work. We all exist in it, so it has an impact on us all.

7

Allies, Unite!

I hope I have convinced you of how universal and dangerous the Infocalypse has become. We are facing a future in which all information is untrustworthy because the environment in which it exists has become so corrupted.

What can an enlightened individual do to make things better rather than worse? In my view, the best use of one's efforts and energies is to contribute to repairing our broken information environment in small but meaningful ways, rather than getting swept up in the polarizing theatre of the Infocalypse and politicking. We need to take the temperature down. When a swimmer is caught in a rip tide, the appropriate survival technique is to swim sideways, not against the current. In a similar way, I believe that the way to resist the Infocalypse is to get outside of it rather than get lost in it. We need to focus on the structure of the information ecosystem rather than the contents of it. To avoid a permanent "fucked-up dystopia" taking hold, we need to understand, defend and then fight back.

ONE: UNDERSTAND

It is not possible to fight something unless there is a general understanding as to what "it" is. That may sound like a simple task, but it is anything but. Even though there are dozens of people and organizations already in this space, there is still no shared conceptual framework and taxonomy to refer to, especially in relation to deepfakes. A clear and consistent conceptual framework needs to be built before these threats can be properly addressed. It is my modest hope that this book can contribute to that aim. I want to reiterate what you urgently need to know:

Our information ecosystem is becoming untrustworthy and dangerous. Society is becoming more familiar with ideas of "disinformation," "misinformation," "conspiracy" and "fake news"—especially in the context of Covid-19. I wanted to find a word that would help me explain that all of these trends are part of a larger, overarching and systematic corruption of our information ecosystem. I settled on the word "Infocalypse," a term coined in 2016 by Avid Ovayda.

The Infocalypse started developing in the last decade and will become more pervasive. While "bad information" is not a new phenomenon, what we are facing now is different in terms of its universality and potency. I cannot pinpoint the exact moment the Infocalypse came into being, but it is certainly a 21st- century phenomenon: an information ecosystem connected to technological revolutions in human connectivity and communication—the Internet, smartphones, social media and video as forms of communication. Now, we face a new and rapidly evolving threat in the form of deepfakes. As I hope to have demonstrated in this book, the crisis of mis- and disinformation in the Infocalypse has real-world consequences that threaten everyone.

Deepfakes are the latest threat. Deepfakes did not emerge in a vacuum, and "fighting them" is part of the larger challenge of protecting the integrity of our information environment. With deepfakes it is especially important to define the taxonomy. I have seen the term "deepfakes" used to mean all synthetic media, to mean face swaps, and to mean face swaps in porn videos. For this reason, I propose that "deepfake" should be used to denote any synthetic media that is used maliciously. There will be many positive-use cases for synthetic media, so this distinction is important. We do not want to throw the baby out with the bathwater.

We still have time to address the deepfake threat. Because the AI behind synthetic media is still nascent, we have time to influence the development of this technology and its outputs. This is the critical time to set standards on how synthetic media is created, labelled and identified. Speed is of the essence. There is a unique opportunity to set standards on how to share and receive AI-generated synthetic media, as well as raise the alarm on its misuses.

TWO: DEFEND

Once we understand the Infocalypse, we can begin to defend against it.

Accurate Information: the first line of defence in the Infocalypse is making sure we have access to accurate information. The Covid-19 pandemic has illustrated how vital this is. Now, more than ever, is the time to support credible journalism and fact-checking organizations. Dozens of such organisations already exist. In the United States they include PolitiFact, which won a Pulitzer Prize for fact-checking during the 2008 election, Snopes and AP Fact Check (from the Associated Press). In Europe, three more of note include the Agence Presse France Fact Check, FullFact and the BBC's Reality Check.

In addition, there are new investigative organizations focusing on disinformation and related information warfare in the Infocalypse. One of these is Bellingcat, which has become world-renowned for its open-source fact-checking work.[1] Founded by Eliot Higgins, it famously established how the Russian military was responsible for shooting down MH-17.

Tools to help protect us from existing mis- and disinformation in our news consumption exist too. Take for example NewsGuard, a company which has developed a browser plug-in that tells you how reliable your news is. For a fee of $2.95 per month, NewsGuard's product provides you with trust ratings for over 4,000 news and information sites. Each trust rating has been graded by trained journalists rather than an algorithm. NewsGuard can tell you who is behind each site, how it's funded, and whether you can trust it.[2] NewsGuard also has a free monthly newsletter, the "Misinformation Monitor," which is an excellent resource.[3] More research is being developed to help newsgathering become more robust in the Infocalypse. Take for example First Draft News. Founded in 2015, it supports journalists and academics with its training courses and Essential Guides series.[4] First Draft is also working with Partnership on AI—

a non-profit organization dedicated to the responsible use of AI—on a research project that will evaluate how to label the synthetic media of the future. While the idea of labelling fake content as "fake" is often touted as a quick and easy solution, there is still a need to establish whether this is a good idea or not. Could it have unexpected effects, by potentially amplifying bad information, for example?[5] Research on future-proofing rigorous journalism in the corrupt information ecosystem is also being undertaken by academics, notably at the Reuters Institute for the Study of Journalism in collaboration with the University of Oxford,[6] the Duke Reporters' Lab[7] and the Nieman Journalism Lab at Harvard.[8]

Humans are vulnerable to what psychologists refer to as the "illusory truth." This is the idea that the longer someone is exposed to something, the more likely he or she is to believe it, even if it is false. To fight our own cognitive biases, mistruths and lies need to be rebutted quickly and repeatedly. Regularly consulting resources from the organizations I mention is the first countermeasure against the Infocalypse. The more you familiarize yourself with mis- and disinformation, the more it works like an inoculation. That means looking at what is happening across the political spectrum, seeing how cheapfakes and deepfakes are being deployed "in the wild."

We all have a personal responsibility here. The more often we engage with these credible resources, the better we become at protecting ourselves, our communities and our societies.

Technical tools: the second line of defence is using the technological tools being built by the "good guys." Technology is merely an amplifier of human intention, and so it is being used for good as well as bad. When I first discovered deepfakes at the end of 2017, I was working with Faculty, an AI company based in London. Faculty had previously built the British government AI software to help detect ISIS propaganda videos.[9] It achieved this by training a machine-learning system with ISIS propaganda videos until the AI was able to "recognize" its hallmarks and detect it when it appeared online.[10]

With Faculty, we were hoping to apply the same principles to detect deepfakes—that is to say, create a machine-learning system that could automatically flag them. At the time, "detection" was still a relatively new field. We only knew of one other serious programme in this area. It was being developed by the Defense Advanced Research Projects Agency (DARPA), the U.S. military agency that researches emerging technologies. DARPA's MediFor aimed to build a platform that would "automatically detect manipulations" and "provide detailed information about how these manipulations were performed."[11]

Since then, the detection field has evolved significantly. In 2018, one of the major problems we faced was that we did not have enough training data to build a detector. (In order to design an AI to recognize deepfakes, you need a lot of deepfakes to train the machine-learning system on.) Since then, the big tech companies have helped open up this field by providing funding and releasing training data for other AI researchers to train their own models. In early 2020, Facebook, Amazon, Microsoft and the Partnership on AI hosted the "Deepfake Detection Challenge," an open competition for deepfake detection tools. The winning submission was to be awarded a prize of $500,000.[12] Google's non-profit arm, Jigsaw, is also developing a host of technological solutions to combat disinformation, including Assembler, an experimental platform that uses new detection technology to fact-check and identify manipulated media— similar to DARPA's MediFor platform.[13] Numerous other firms are developing their own detection technologies—this is one part of what DeepTrace does. Another notable effort is being led by The AI Foundation, which has developed a detection software product it calls "Reality Defender."

As the quality of synthetic media improves, we humans

won't be able to detect deepfakes, which is why it makes sense to invest in AI detection tools. However, this approach has its own pitfalls. Deepfake detection is a constantly evolving game of cat-and-mouse. As AI detection tools get better, so too will deepfakes. Theoretically, it is possible that synthetic-media generation will get so good that detection will be impossible. The jury is still out on whether or not we will get to that point.

Aside from "detection," technologists are also working on provenance—or proving the veracity of authentic media. WITNESS, the human rights organization mentioned in Chapter 4, has developed an app called ProofMode which verifies media as authentic at its point of capture. For WITNESS, this tool is an essential part of the work of activists documenting human-rights abuses in dangerous parts of the world. Another company I spoke to, TruePic, has a similar tool. It uses its own patented "controlled capture" software to verify the origins and metadata of photos and videos. Mounir Ibrahim, TruePic's Vice-President of Strategic Indicatives, told me that this is particularly important in the private sector for industries like insurance. Without technology to verify the authenticity of claims, the insurance industry could collapse under a tsunami of fraudulent claims based on fake visual evidence. Mounir told me that Truepic would like to go further, however, and is examining the possibility of developing hardware that would authenticate media even closer to its point of capture (that is to say, by implanting it directly into a mobile phone, so that any media recorded with that device has permanent and distinct markers to show where and when it was captured.)

A project at the New York University Tandon School of Engineering is based on a similar idea, trying to embed hardware directly into a camera that places watermarks on each photograph's code. These watermarks will provide proof as to a photograph's authenticity to forensics analysts. With funding from IBM, *The New York Times'* News Provenance Project is working with publishers and platforms to develop ways to illustrate the provenance of visual content online.[14] One idea being explored is the use of blockchain as a potentially secure method of proving where photographs come from, so they cannot be miscontextualized as cheapfakes.[15] AI research is also bolstering this line of defence. The research programmes at the University of Washington, Stanford, MIT, Carnegie Mellon, the University of Southern California and the Technical University of Munich are especially notable in this regard.

Society-wide responses: broader societal approaches to the Infocalypse through public policy are probably the least developed, and most complicated, line of defence. In part, this is due to the fact that there needs to be a more robust debate about the trade-offs between security, liberty and privacy as we move forward. The question at the heart of this is who should have the power to decide what information is good or bad, and how can that be adjudicated without bias? It is of paramount importance to get this balance right. This is especially difficult in Western democracies where free speech and freedom of information will become increasingly difficult to reconcile with the threats posed by mis- and disinformation.

An example of how complicated this is can be seen in the on-going battle between President Donald Trump and Twitter. When Twitter for the first time put a health warning on tweets from President Trump as "potentially misleading" in May 2020, Trump escalated the matter into a war over the First Amendment right to free speech. The tweets in question were the ones in which the President makes the claim of a "rigged election" and voter fraud by mail-in ballots that I covered in Chapter 3. Objectively, Trump's statements were misleading. American electoral history shows that there is very little chance of this type of voter fraud happening, let alone that it is being organized as a mass exercise by Democrats.

Far from undermining the First Amendment, constitutional experts say Twitter acted within its rights. First, because the First Amendment protects private companies from government interference, and Twitter can set its own terms and policies, regardless of what the President wants it to do. And secondly, because Twitter did not ban Trump or remove his posts, but merely flagged his comments and added links with more information, it is protected as counterspeech. That didn't stop the President from signing an executive order against Twitter, claiming that he was doing it to "defend free speech from one of the gravest dangers it has faced in American history."[16] Legal experts say that Trump is setting a dangerous legal precedent by trying to pressurize companies into giving his content preferential treatment.[17] So even when something is demonstrably false (as in this case), its labelling has become partisan. One poll shows that 77 per cent of Republican voters agree with Trump's often repeated and unsubstantiated claim that the big tech companies purposefully suppress conservative views and, therefore, free speech.

There is no point in pretending that there are easy policy solutions to defend against the Infocalypse. The challenges are complicated, and vital work is being done to try and unpick some of these questions. There are numerous organizations to engage with in this space, such as activists at the Electronic Frontier Foundation, Avaaz, PEN America, Access Now and WITNESS. Others, like the Center for Humane Technology, approach these issues from a consumer-rights angle, working with regulators and tech companies to examine how the business model of the big tech platforms may contribute to "internet addiction, political extremism and misinformation." Other bodies like the Disinfo Portal at the Atlantic Council and the EU's High-Level Expert Group on Fake News and Online Disinformation are aiming to formulate effective responses to foreign state-led attacks. The "EUvsDinfo" website is an excellent resource on Russian disinformation activities in particular. Academics are also contributing to this field, notably the Stanford Internet Observatory and the Cyber Policy Center at Stanford; the Information Disorder Lab at Harvard, the Oxford Internet Institute and the Center for Media Engagement at the University of Texas among others.

THREE: FIGHT

The final piece of the puzzle to resist the "fucked-up dystopia" is fighting back: a more proactive rather than reactive approach. Of course, no one can on take this challenge alone. We will need alliances of the willing. Take the example of Operation Double Deceit, the Russian-influenced operation in Ghana. This was exposed by the joint efforts of CNN, Graphika, Facebook and Twitter. These alliances would benefit from a more permanent architecture to incentivize collaboration rather than being pulled together for ad hoc projects.

One organization I encountered, Deep Trust Alliance, was set up with the purpose of trying to build such alliances.[18] Its CEO and founder, Kathryn Harrison, told me that her long career at IBM had taught her that solving complex problems such as disinformation and deepfakes needs collaboration between the policy, tech and business communities.

She admitted that this approach can be tough to coordinate. "Humans don't always play nicely together in general [so it] is difficult to get people to come together to collaborate and come to a single point of view." It is her aim, however, to "connect the dots."[19] Even if the work is slow, and the victories small at first, Kathryn is convinced it is worth it. As she added with a laugh, "I am an optimist. If I wasn't, I wouldn't be doing this work [...] We need to be happy with incremental success while shooting for the North Star."

Kathryn is right. The dots need to be connected, and even incremental success is a start. When it comes to proactively fighting the Infocalypse, there are some hopeful precedents.

Take, for example, the case of the small Baltic nation of Estonia, with a population of 1.3 million. Traditionally a target for Soviet information operations, it learned how to fight back. When Estonia gained its independence in 1991 after five decades of Soviet occupation, it quickly moved away from Moscow's sphere of influence through alliances with NATO and the United States. However, even after the collapse of the Soviet Union, Russia continued to treat Estonia as though it were a Soviet satellite state.

In 2007, the Estonian government decided to relocate the "Bronze Soldiers of Talinn," a memorial to the occupying Soviet soldiers killed in the Second World War. The Kremlin condemned the plan, but the Estonian government went ahead anyway. Soon after, Estonia was hit with a series of vicious cyber-attacks against its government, media and banking infrastructure. The attacks lasted over three weeks. Russia denied it was behind the assault, but Estonian officials had no doubt the attacks were coming from Moscow, as a retaliation for the movement of the statue.

Instead of crumbling, Talinn used the lesson to reinforce its defences and fight back. It did this by taking a society-wide approach that engaged all its citizens. First, it assessed where Russia might attack next and launched an early-warning system for debunking Russian disinformation in national discourse. For example, through its "Baltic Elves"—volunteers who monitor the Internet for Russian disinformation. Secondly, it started building up its digital defences, through the Cyber Defence League, an army of volunteer IT and disinformation specialists who focused on sharing threat information and preparing society for responding to cyber incidents. Today, Estonia is one of the most sophisticated digital nations in the world: 99 per cent of public services are online and nearly one third of citizens vote online, and it has not suffered a further Russian breach of its defences. Thirdly, it mobilized its entire society against the threat. In 2010, Estonia introduced a new long-term national defence strategy that emphasized "psychological defence," which it defined as the "development, preservation and protection of common values associated with social cohesion and the sense of security."[20]

I spoke to Johannes Tammekänd, the CEO and founder of Sentinel, a cybersecurity, AI and policy group working with the Estonian government on the emerging deepfake threat. Discussing why Estonia has been so successful in the face of Russian disinformation, Johannes pointed to its "multi-layer" defences. "It has to be like a medieval fortress," he explained. "First a moat, then the outer walls, then the inner walls." Estonia has been good at joining the dots. And it did this by enhancing cooperation domestically and with its NATO allies to raise awareness and prepare defences. As Johannes told me, "You can fool some people some of the time, but you can't fool all the people all of the time." Ultimately, he concludes that it is a question for society as to whether or not it wants to exist in a corrupt information ecosystem. Estonia's defence is predicated on the notion that once society knew about the dangers it was facing, it said "no." It "understood where all this is going" and put in the hard effort to fight back.

HOPE FOR THE FUTURE

We are entering a new age in human development. The way in which we communicate as a species is being radically transformed. One side effect is that we now exist in an increasingly dangerous and untrustworthy information ecosystem. However, there is hope. The forces to fight the Infocalypse are already coming together and growing in strength. They are helping us understand the threat, and they have also started building the solutions and alliances to help safeguard us all. But they need our support. We can all help by sharing this understanding, preparing our defences and fighting back. Time is of the essence. If you do not want the "fucked-up dystopia" to become a permanent reality, engage now. Be careful about what information you share. Verify your sources. Correct yourself when you get something wrong. Be wary of your own political biases. Be sceptical, but not cynical. If you want to learn more, start with all the organizations I have listed as a resource at the end of this book. It is time for all allies to unite. As the Estonians did to the Russians, we still have the chance to say "no" to the Infocalypse.

RESOURCES

Fact-checking organizations
- APF Fact Check—factcheck.afp.com
- AP Fact—apnews.com/APFactCheck
- BBC Reality Check—bbc.co.uk/news/reality_check
- FullFact—fullfact.org
- Politfact—politifact.com
- Snopes—snopes.com

Media provenance
- ˮContent Authenticity Initiative (Adobe)—contentauthenticity.org
- Digimac—digimap.edina.ac.uk
- News Provenance Project—newsprovenanceproject.com
- Pressland—pressland.com

Disinformation detection and protection
- Amped—ampedsoftware.com
- AI Foundation—aifoundation.com
- Bellingcat—bellingcat.com
- DARPA—darpa.mil
- EUvsDisinfo—euvsdisinfo.eu
- The Citizen Lab at the University of Toronto—citizenlab.ca
- DeepTrace—deeptracelabs.com
- Jigsaw—jigsaw.google.com
- NewsGuard—newsguardtech.com
- Truepic—truepic.com

Social-media analysis
- Botswatch—botswatch.io
- Dataminr—dataminr.com
- Graphika—graphika.com
- Storyful—storyful.com

Best practice (media)
- Duke Reporters' Lab—reporterslab.org
- Credibility Coalition—credibilitycoalition.org
- First Draft News—firstdraftnews.org
- News Literacy Project—newslit.org
- News Integrity Initiative, Newmark School of
  Journalism, The City University of New York—
  journalism.cuny.edu/centers/tow-knight-center-
  entrepreneurial-journalism/news-integrity-initiative/
- Nieman Lab, Harvard University—niemanlab.org
- Partnership on AI—partnershiponai.org
- Reuters Institute—reutersinstitute.politics.ox.ac.uk

Policy/society
- Access Now—accessnow.org
- Alliance for Securing Democracy—
  securingdemocracy.gmfus.org
- Anti-Defamation League—adl.org
- Center for Humane Technology—
  humanetech.com/problem/
- Center for Media Engagement, Moody College
  of Communication, University of Texas at Austin—
  mediaengagement.org/
- Cyber Policy Center, Stanford University—
  cyber.fsi.stanford.edu

- Data and Society, Disinformation Action Lab—datasociety.net/research/disinformation-action-lab/
- DeepTrust Alliance—deeptrustalliance.org
- Digital Forensics Research Lab and DisinfoPortal, Atlantic Council—atlanticcouncil.org/programs/digital-forensic-research-lab/
- Electronic Frontier Foundation—eff.org
- Information Disorder Lab, Shorenstein Centre, Harvard University—shorensteincenter.org/about-us/areas-of-focus/misinformation/
- Internet Observatory, Stanford University—cyber.fsi.stanford.edu/io/content/io-landing-page-2
- OpenAI—openai.com
- PEN America—pen.org
- Partnership on AI—partnernshiponai.org
- The Truthiness Collaboration, Annenberg Innovation Lab, University of Southern California—annenberglab.com
- Wikimedia—wikimedia.org
- WITNESS—witness.org

ENDNOTES

Introduction: "fucked-up dystopia"

1   www.youtube.com/watch?time_continue=36&v=cQ54GDm1eL0&feature=emb_logo

2   www.youtube.com/playlist?list=PLrRNOOLd8VZc0EOdyukjieeuhf1vFVPr_

3   www.bellingcat.com/news/uk-and-europe/2015/10/08/mh17-the-open-source-evidence/

4   Intelligence and Security Committee of Parliament, *Annual Report, 2016–2017* (HMSO, 2017), p 52, available at https://sites.google.com/a/independent.gov.uk/isc/files/2016-2017_ISC_AR.pdf?attredirects=1

5   Digital Forensic Research Lab, "Question That: RT's Military Mission," https://medium.com/dfrlab/question-that-rts-military-mission-4c4bd9f72c88

6   Darrell Etherington, "People now watch 1 billion hours of YouTube hours per day," techcrunch.com, 28 February 2017, https://techcrunch.com/2017/02/28/people-now-watch-1-billion-hours-of-youtube-per-day/

7   'More than 90%' of Russian airstrikes in Syria have not targeted Isis, US says," *The Guardian*, 7 October 2015, available at www.theguardian.com/world/2015/oct/07/russia-airstrikes-syria-not-targetting-isis

8   Lizzie Dearden, "Russia and Syria 'weaponising' refugee crisis to destabilise Europe, Nato commando claims," *The Independent*, 3 March 2016, available at www.independent.co.uk/news/world/middle-east/russia-and-syria-weaponising-refugee-crisis-to-destabilise-europe-nato-commander-claims-a6909241.html

9   Henri Neurendorf, "Ai Weiwei commemorates drowned refugees with public installation during Berlin Film Festival," *artnet news*, 15 February 2016, https://news.artnet.com/art-world/ai-weiwei-life-jackets-installation-berlin-427247

10 Todd Bensen, "What terrorist migration over European borders can teach about American border security," Report for the Center for Immigration Studies, 6 November 2019, available at https://cis.org/Report/Terrorist-Migration-Over-European-Borders

11 Stefan Meister, "The 'Lisa case': Germany as a target of Russian disinformation," *Nato Review*, 25 July 2016, www.nato.int/docu/review/articles/2016/07/25/the-lisa-case-germany-as-a-target-of-russian-disinformation/index.html

12 Ben Knight, "Teenage girl admits making up migrant rape claim that outraged Germany," *The Guardian*, 31 January 2016, available at www.theguardian.com/world/2016/jan/31/teenage-girl-made-up-migrant-claim-that-caused-uproar-in-germany

13 Adrienne Klasa, Valerie Hopkins, Guy Chazan, Henry Foy and Miles Johnson, "Russia's long arm reaches to the right in Europe," *Financial Times*, 23 May 2019, available at www.ft.com/content/48c4bfa6-7ca2-11e9-81d2-f785092ab560

14 BBC News, "Russian hackers 'target' presidential candidate Macron," www.bbc.co.uk/news/technology-39705062

15 www.voteleavetakecontrol.org/briefing_immigration.htm

Chapter 1: R/Deepfakes

1   Michael Waters, "The great lengths taken to make Abraham Lincoln look good in portraits," *Atlas Obscura*, 12 July 2017, www.atlasobscura.com/articles/abraham-lincoln-photos-edited

2   Peter Eaves, "Traces of human tragedy: the David King collection," Tate Research Feature, July 2018, www.tate.org.uk/research/features/human-tragedy-david-king-collection

3   Ibid.

4   Eryn J Newman et al., "Truthiness and falsiness of trivia claims depend on judgmental contexts," https://publications.aston.ac.uk/id/eprint/25450/1/Truthiness_and_falsiness_of_trivia_claims_depend_on_judgmental_contexts.pdf

5   Chris Evangelista, "Martin Scorsese is obsessing over the 'youthification' CGI in 'The Irishman,'" www.slashfilm.com/the-irishman-cgi/

6   Matt Miller, "Some deepfaker on YouTube spent seven days fixing the shitty de-aging in *The Irishman*," *Esquire*, 7 January 2020, available at www.esquire.com/entertainment/movies/a30432647/deepfake-youtube-video-fixes-the-irishman-de-aging/

7   www.youtube.com/watch?time_continue=238&v=dyRvbFhknRc&feature=emb_logo

8   Samantha Cole, "AI-assisted fake porn is here and we're all fucked," vice.com, 11 December 2017, www.vice.com/en_us/article/gydydm/gal-gadot-fake-ai-porn

9   www.youtube.com/watch?v=FqzE6NOTMOg

10   www.youtube.com/watch?v=2svOtXaD3gg&t=194s

11   www.youtube.com/watch?v=_OqMkZNHWPo&feature=emb_title

12   Giorgio Patrini, "Mapping the deepfake landscape," deeptracelabs.com, 7 October 2019, https://deeptracelabs.com/mapping-the-deepfake-landscape/

13   www.adultdeepfakes.com

14   Drew Harwell, "Scarlett Johansson on fake AI-generated sex videos: 'Nothing can stop someone from cutting and pasting my image,'" *Washington Post*, 31 December 2018, available at www.washingtonpost.com/technology/2018/12/31/scarlett-johansson-fake-ai-generated-sex-videos-nothing-can-stop-someone-cutting-pasting-my-image/

15   www.youtube.com/watch?v=BxIPCLRfk8U

16   "AI creates fashion models with custom outfits and poses," *Synced*, 29 August 2019, https://syncedreview.com/2019/08/29/ai-creates-fashion-models-with-custom-outfits-and-poses/

17   www.youtube.com/watch?v=FzOVqClci_s

18   Tiffany Hsu, "An ESPN commercial hints at advertising's deepfake future," *The New York Times*, 22 April 2020, available at www.nytimes.com/2020/04/22/business/media/espn-kenny-mayne-state-farm-commercial.html

Chapter 2: Russia: the master

1   www.youtube.com/watch?v=bX3EZCVj2XA

2   Charles Dervarics, "Conspiracy beliefs may be hindering HIV prevention among African Americans," prb.org, 1 February 2005, www.prb.org/conspiracybeliefsmaybehindering-hivpreventionamongafricanamericans/

3   Reuters, "Factbox: U.S. intel report on Russian cyber attacks in 2016 election," 6 January 2016, www.reuters.com/article/us-usa-russia-cyber-intel-factbox/factbox-u-s-intel-report-on-russian-cyber-attacks-in-2016-election-idUSKBN14Q2HH

4   Ibid.

5   Special Counsel Robert S Mueller, III, *Report on the Investigation Into Russian Interference in the 2016 Presidential Election* , vol. 1 (U.S. Department of Justice, 2019), available at www.justice.gov/storage/report.pdf

6   Ibid., p 14.

7   Ibid., pp 24–5.

8   Department of Defense and Joint Chief of Staff, *Russian Strategic Intentions*, A Strategic Multilayer Assessment (SMA) White Paper, May 2019, available at www.politico.com/f/?id=0000016b-a5a1-d241-adff-fdf908e00001

9   https://intelligence.house.gov/social-media-content/social-media-advertisements.htm

10 Politico Staff, "The social media ads Russia wanted Americans to see," *Politico*, 1 November 2017, www.politico.com/story/2017/11/01/social-media-ads-russia-wanted-americans-to-see-244423

11 Mueller, *Report on the Investigation into Russian Interference in the 2016 Presidential Election*, p 29.

12 Ibid., p 31.

13 Ali Breland, "Thousands attended protest organized by Russians on Facebook," *The Hill*, 31 October 2017, https://thehill.com/policy/technology/358025-thousands-attended-protest-organized-by-russians-on-facebook

14 http://web.archive.org/web/20161113035441/https://www.facebook.com/events/535931469910916/

15 Sam Harris, "#145—The information war; A Conversation with Renée DiResta," podcast, 2 January 2019, https://samharris.org/podcasts/145-information-war/

16 Josh Hafner, "Army of Jesus"? How Russia messed with Americans online, *USA Today*, 15 December 2019, available at https://eu.usatoday.com/story/news/politics/onpolitics/2017/11/01/onpolitics-today-army-jesus-how-russia-messed-americans-online/823842001/

17 Ibid.

18 Joan Donavan, "How memes got weaponized: a short history," *MIT Technology Review*, 24 October 2019, available at www.technologyreview.com/s/614572/political-war-memes-disinformation/

19 Philip N Howard, Bharath Ganesh, Dimitra Liotsiou, John Kelly and Camille François, "The IRA, Social media and political polarization in the United States, 2012–2018," Working Paper 2018.2. Oxford, UK: Project on Computational Propaganda. comprop.oii.ox.ac.uk

20 https://graphika.com/uploads/Graphika_Report_IRA_in_Ghana_Double_Deceit.pdf

21 Ibid.

22 Ibid.

23 Ibid.

24 Ibid.

25 Howard et al., "The IRA, social media and political polarization in the United States, 2012–2018."

26 Bryan Bender, "Russia beating U.S. in race for global influence, Pentagon study says," *Politico*, 30 June 2019, available at www.politico.com/story/2019/06/30/pentagon-russia-influence-putin-trump-1535243?nname=playbook&nid=0000014f-1646-d88f-a1cf-5f46b7bd0000&nrid=0000016a-d0df-db42-ad6e-fedfc07f0000&nlid=630318

27 Andrew Desiderio and Kyle Cheney, "Mueller refutes Trump's 'no collusion, no obstruction' line," politico.eu, 24 July 2019, www.politico.eu/article/mueller-refutes-trumps-no-collusion-no-obstruction-line/

28 www.youtube.com/watch?v=LiaMludqL1A

29 Bryan Bender, "Russia beating U.S. in race for global influence, Pentagon study says," see note 26 above.

30 Samantha Bradshaw and Philip N Howard, "The global disinformation order: 2019 global inventory of organised social media manipulation," Working Paper 2019.3. Oxford, UK: Project on Computational Propaganda. comprop.oii.ox.ac.uk

31 Jacon N Shapiro, "Trends in online foreign influence efforts," *Empirical Studies of Conflict*, 2019, available at https://esoc.princeton.edu/files/trends-online-foreign-influence-efforts

Chapter 3: The West: the internal threat

1 Freedom House, *Freedom in the World 2020*, https://freedomhouse.org/report/freedom-world/2020/leaderless-struggle-democracy

2 www.allianceofdemocracies.org/wp-content/uploads/2018/06/Democracy-Perception-Index-2018-1.pdf

3 Meghan Keneally, "Donald Trump's history of raising birther questions about President Obama," ABC News, 18 September 2015, https://abcnews.go.com/Politics/donald-trumps-history-raising-birther-questions-president-obama/story?id=33861832

4 Glenn Kessler and Scott Clement, "Trump routinely says things that aren't true. Few Americans believe him," *Washington* Post, 14 December 2018, available at www.washingtonpost.com/graphics/2018/politics/political-knowledge-poll-trump-falsehoods/

5 "Public trust in government, 1958–2019," Pew Research Center, 11 April 2019, www.people-press.org/2019/04/11/public-trust-in-government-1958-2019/

6 A line from Trump's 2015 candidacy speech; full transcript available at https://time.com/3923128/donald-trump-announcement-speech/

7 www.kff.org/coronavirus-covid-19/report/kff-health-tracking-poll-may-2020/

8   Michael M Grynbaum, "Trump's briefings are a ratings hit. Should networks cover them
    live?," *The New York Times*, 25 Match 2020, available at www.nytimes.com/2020/03/25/
    business/media/trump-coronavirus-briefings-ratings.html

9   Matthias Lüfkens, "Hillary Clinton v Donald Trump: who's winning on Twitter?," weforum.
    com, 2 August 2016, www.weforum.org/agenda/2016/08/hillary-clinton-or-donald-
    trump-winning-on-twitter/

10 https://twitter.com/realdonaldtrump/status/949618475877765120?lang=en

11 https://twitter.com/realdonaldtrump/status/1021234525626609666?lang=en

12 Manuela Tobias, "Fact-checking distorted video Sarah Sanders used to bar a CNN White
    House reporter," *Politifact*, 8 November 2018, www.politifact.com/article/2018/nov/08/
    fact-checking-misleading-video-sarah-sanders-used-/

13 Joanna Walters, "Jim Acosta: White House backs down in fight over CNN reporter's pass,"
    The Guardian, 19 November 2018, available at www.theguardian.com/
    media/2018/nov/19/jim-acosta-white-house-press-pass-trump-administration-suspend-letter

14 Makena Kelly, "Trump tests disinformation policies with new Pelosi video," *The Verge*, 7
    February 2020, www.theverge.com/2020/2/7/21128317/nancy-pelosi-
    donald-trump-disinformation-policy-video-state-of-the-untion

15 Fact Check 17856, "The Fact Checker's ongoing database of the false or misleading
    claims made by President Trump since assuming office," *Washington* Post, 29 May
    2020, available at http://wapo.st/trumpclaimsdb?claim=18486

16 https://twitter.com/SilERabbit/status/1254551597465518082?s=20

17 Katherine Schaeffer, "Far more Americans see 'very strong' partisan conflicts now than
    in the last two presidential election years," *Facttank*, 4 March 2020, www.pewresearch.
    org/fact-tank/2020/03/04/far-more-americans-see-very-strong-partisan-conflicts-now-
    than-in-the-last-two-presidential-election-years/

18 See Jennifer Mercieca, *Demagogue for President: The Rhetorical Genius of Donald
    Trump* (Texas A&M University Press, 2020).

19 Shayanne Gal and Mariana Alfaro, "30 of Trump's most famous quotes since becoming
    president," *Business Insider*, 11 January 2019, available at www.businessinsider.com/
    trump-quotes-since-becoming-president-2018-6?r=US&IR=T

20 Fact Check 17639, "The Fact Checker's ongoing database of the false or misleading
    claims made by President Trump since assuming office," *Washington* Post, 29 May
    2020, available at http://wapo.st/trumpclaimsdb?claim=17939

21 https://twitter.com/realDonaldTrump/status/1265011145879977985

22 https://twitter.com/realDonaldTrump/status/1265608389905784834

23 Oliver Darcy, "Trump viciously attacks NBC News reporter in extended rant after being
    asked for message to Americans worried about coronavirus," CNN News, 21 March
    2020, https://edition.cnn.com/2020/03/20/media/trump-rant-at-nbc-news-peter-
    alexander/index.html

24 Morgan Chalfant, "White House defends Trump's 'human scum' remark," *The Hill*, 24
    October 2019, https://thehill.com/homenews/administration/467260-white-house-
    defends-trumps-human-scum-remark

25 "George Floyd death homicide, official post-mortem declares," BBC News, 2 June 2020, www.bbc.co.uk/news/world-us-canada-52886593

26 Monica Rhor, "Rhor: George Floyd's last words speak truth of black life in America. Are we listening?" *Houston Chronicle*, 28 May 2020, available at www.houstonchronicle.com/opinion/outlook/article/Rhor-Floyd-s-last-words-speak-truth-of-black-15302050.php

27 https://twitter.com/joshscampbell/status/1266805337652449283

28 https://twitter.com/realDonaldTrump/status/1266914470066036736?s=20

29 Michael S. Rosenwald, " 'When the looting starts, the shooting starts': Trump quotes Miami police chief's notorious 1967 warning," *Washington Post*, 29 May 2020, available at www.washingtonpost.com/history/2020/05/29/when-the-looting-starts-the-shooting-starts-trump-walter-headley/

30 https://twitter.com/realDonaldTrump/status/1266711223657205763

31 https://twitter.com/realDonaldTrump/status/1266799941273350145

32 https://twitter.com/realDonaldTrump/status/1266799941273350145

33 Matt Zapotosky, "Trump threatens military action to quell protests, and the law would let him do it," *Washington* Post, 2 June 2020, available at www.washingtonpost.com/national-security/can-trump-use-military-to-stop-protests-insurrection-act/2020/06/01/c3724380-a46b-11ea-b473-04905b1af82b_story.html

34 Ibid.

35 Craig Timberg, Tony Romm, Aaron C. Davis and Elizabeth Dwoskin, "Secret campaign to use Russian-inspired tactics in 2017 Ala. Election stirs anxiety for Democrats," *Washington* Post, 6 June 2019, available at www.washingtonpost.com/business/technology/secret-campaign-to-use-russian-inspired-tactics-in-2017-alabama-election-stirs-anxiety-for-democrats/2019/01/06/58803f26-0400-11e9-8186-4ec26a485713_story.html

36 www.newsguardtech.com/misinformation-monitor-may-2020/

37 https://projects.fivethirtyeight.com/polls/

Chapter 4: The Rest: global information disorder

1 www.statista.com/statistics/883751/myanmar-social-media-penetration/

2 Timothy McLaughlin, "How Facebook's rise fueled chaos and confusion in Myanmar," *Wired*, 7 June 2018, available at www.wired.com/story/how-facebooks-rise-fueled-chaos-and-confusion-in-myanmar/

3 www.hrw.org/tag/rohingya-crisis

4 Julia Carrie Wong, " 'Overreacting to failure': Facebook's new Myanmar strategy baffles local activists," *The Guardian*, 7 February 2019, available at www.theguardian.com/technology/2019/feb/07/facebook-myanmar-genocide-violence-hate-speech

5 Elyse Samuels, "How misinformation on WhatsApp led to a mob killing in India," *Washington Post*, 21 February 2020, available at www.washingtonpost.com/politics/2020/02/21/how-misinformation-whatsapp-led-deathly-mob-lynching-india/

6 www.witness.org/witness-deepfakes-prepare-yourself-now-report-launched/

7 Corin Faife, "In Africa, fear of state violence informs deepfake threat," WITNESS blog, 9 December 2019, https://blog.witness.org/2019/12/africa-fear-state-violence-in-forms-deepfake-threat/

8 Rana Ayyub, *Gujarat Files: Anatomy of a Cover-up* (CreateSpace, 2016).

9 www.theguardian.com/commentisfree/2012/mar/14/new-india-gujarat-massacre

10 Rana Ayyub, "I was the victim of a deepfake porn plot intended to silence me," blog, *HuffPost*, UK Edition, www.huffingtonpost.co.uk/entry/deepfake-porn_uk_5bf2c126e4b-0f32bd58ba316

11 Ibid.

12 Ibid.

13 Leanne de Bassompierre, "Gabon's President Bongo had a stroke, AFP says, citing Moussavou," bloomberg.com, 9 December 2018, www.bloomberg.com/news/articles/2018-12-09/gabon-s-president-bongo-had-a-stroke-afp-says-citing-moussavou

14 www.youtube.com/watch?v=F5vzKs4z1dc

15 Sarah Cahian, "How misinformation helped spark an attempted coup in Gabon," *Washington Post*, 13 February 2020, available at www.washingtonpost.com/politics/2020/02/13/how-sick-president-suspect-video-helped-sparked-an-attempted-coup-gabon/

16 Ali Breland, "The bizarre and terrifying case of the 'deepfake' video that helped bring an African nation to the brink," *MotherJones*, 15 March 2019, www.motherjones.com/politics/2019/03/deepfake-gabon-ali-bongo/

17 Sarah Cahian, "How misinformation helped spark an attempted coup in Gabon," *Washington Post*, 13 February 2020, available at www.washingtonpost.com/politics/2020/02/13/how-sick-president-suspect-video-helped-sparked-an-attempted-coup-gabon/

18 www.youtube.com/watch?v=F5vzKs4z1dc

19 Tashny Sukumaran, "Malaysia's Azmin Ali sex scandal: minister's aide asked me to lie, says Haziq Abdul Aziz, who claims he is the other man in viral video," scmp.com, 13 June 2019, www.scmp.com/week-asia/politics/article/3014355/malaysias-azmin-ali-sex-scandal-ministers-aide-asked-me-lie-says

20 Philip Golingai, "Is it Azmin or a deepfake?," *The Star*, 15 June 2019, available at www.thestar.com.my/opinion/columnists/one-mans-meat/2019/06/15/is-it-azmin-or-a-deep-fake

Chapter 5: Deepfakes in the Wild

1 Harriet Johnston, "'Vulnerable' Prince Harry 'stands by' what he said in phone calls with Russian pranksters posing as Greta Thunberg and her father, but still feels 'violated,' royal expert claims," *Daily Mail*, 17 March 2020, available online at www.dailymail.co.uk/femail/article-8120967/Prince-Harry-stands-said-prank-phone-calls-felt-violated-royal-expert-claims.html

2 Jonathan Shen and Ruoming Pang, "Tacotron 2: generating human-like speech from text," Google AI Blog, 19 December 2017, https://ai.googleblog.com/2017/12/tacotron-2-generating-human-like-speech.html

3 https://knowyourmeme.com/memes/navy-seal-copypasta

4 www.youtube.com/watch?v=zBUDyntqcUY

5 www.youtube.com/watch?v=drirw-XvzzQ

6 www.youtube.com/watch?v=vk89hEhst88

7 Matt Novak, "Make Jordan Peterson say anything you want with this spooky audio generator," Gizmodo, 16 August 2019, https://gizmodo.com/make-jordan-peterson-say-anything-you-want-with-this-sp-1837306431

8 Valerie Solanas, "The SCUM Manifesto," first published 1968; available at https://www.ccs.neu.edu/home/shivers/rants/scum.html

9 Jordan Peterson, "The deepfake artists must be stopped before we no longer know what's real," National Post, 23 August 2019, available at https://nationalpost.com/opinion/jordan-peterson-deep-fake

10 Ibid.

11 Catherine Stupp, "Fraudsters used AI to mimic CEO's voice in unusual cybercrime case," The Wall Street Journal, 30 August 2019, available at www.wsj.com/articles/fraudsters-use-ai-to-mimic-ceos-voice-in-unusual-cybercrime-case-11567157402

12 www.crowe.com/global/news/fraud-costs-the-global-economy-over-us$5-trillion

13 https://www.descript.com/lyrebird-ai

14 https://s3.amazonaws.com/media.mediapost.com/uploads/EconomicCostOfFakeNews.pf

15 This made Tesla more valuable than the next two biggest American car companies (Ford and General Motors) combined. Both Ford and GM delivered at least 2 million cars in the U.S. in 2019, whereas Tesla had delivered only 360,000.

16 Richard Henderson, "Tesla short sellers take further hit in battle with Elon Musk," Financial Times, 3 February 2020, available at www.ft.com/content/32c9c8c4-4478-11ea-a43a-c4b328d9061c

17 Ibid.

18 Russell Hotten, "Elon Musk tweet wipes $14bn off Tesla's value," BBC News, 1 May 2020, www.bbc.co.uk/news/business-52504187

19 Craig Silverman, Jane Lytvynenko and William Kung, "Disinformation for hire: how a new breed of PR firms is selling lies online," BuzzFeed News, 6 January 2020, www.buzzfeednews.com/article/craigsilverman/disinformation-for-hire-black-pr-firms

20 www.ted.com/speakers/noelle_martin

21 Samantha Cole, "This horrifying app undresses a photo of any woman with a single click," Vice, 26 June 2019, www.vice.com/en_us/article/kzm59x/deepnude-app-creates-fake-nudes-of-any-woman

22 Giorgio Patrini, "Mapping the deepfake landscape," deeptracelabs.com, 7 October 2019, https://deeptracelabs.com/mapping-the-deepfake-landscape/

23 Ian Sample, "Internet 'is not working for women and girls,' says Berners-Lee," *The Guardian*, 12 March 2020, www.theguardian.com/global/2020/mar/12/internet-not-working-women-girls-tim-berners-lee

24 https://wikileaks.org/podesta-emails/press-release

25 https://dcpizzagate.wordpress.com/2016/11/07/first-blog-post/

26 https://dcpizzagate.wordpress.com/2016/11/07/first-blog-post/

27 Burt Helm, "Pizzagate nearly destroyed my restaurant. Then my customers helped me fight back," *Inc.*, July/August 2017, available at www.inc.com/magazine/201707/burt-helm/how-i-did-it-james-alefantis-comet-ping-pong.html

28 Ibid.

Chapter 6: Covid 19: a global virus

1 Veronika Melkozerova and Oksana Parafeniuk, "How coronavirus disinformation caused chaos in a small Ukrainian town," NBC News, 3 March 2020, www.nbcnews.com/news/world/how-coronavirus-disinformation-caused-chaos-small-ukrainian-town-n1146936

2 Andrew Green, "Obituary: Li Wenliang," *The Lancet*, 395.10225 (18 February 2020), available at wwww.thelancet.com/journals/lancet/article/PIIS0140-6736(20)30382-2/fulltext

3 "He warned of coronavirus. Here's what he told us before he died," *The New York Times*, 7 February 2020, available at www.nytimes.com/2020/02/07/world/asia/Li-Wen-liang-china-coronavirus.html

4 "Chinese 'disinformation' on coronavirus costing lives, say MPs," *PoliticsHome*, 6 April 2020, www.politicshome.com/news/article/chinese-disinformation-on-coronavirus-cost-ing-lives-say-mps

5 Ibid.

6 www.facebook.com/XinhuaUK/posts/1594877697326136?__tn__=-R

7 "Chinese 'disinformation' on coronavirus costing lives, say MPs," *PoliticsHome*, 6 April 2020, www.politicshome.com/news/article/chinese-disinformation-on-coronavirus-costing-lives-say-mps

8 Julian E Barnes, Matthew Rosenberg and Edward Wong, "As virus spreads, China and Russia see openings for disinformation," *The New York Times*, 28 March 2020, available at www.nytimes.com/2020/03/28/us/politics/china-russia-coronavirus-disinformation.html

9 Agence France-Presse, "'Might be US Army who brought virus epidemic to Wuhan': Chinese official," ndtv.com, 13 March 2020, www.ndtv.com/world-news/us-army-may-have-brought-coronavirus-into-china-claims-official-zhao-lijian-2194096

10 Oliver Milman, "Trump administration cut pandemic early warning program in September," 3 April 2020, *The Guardian*, available at www.theguardian.com/world/2020/apr/03/trump-scrapped-pandemic-early-warning-program-system-before-coronavirus

11 Luciana Borio and Scott Gottlieb, "Stop a U.S. coronavirus outbreak before it starts," *The Wall Street Journal*, 4 February 2020, available at www.wsj.com/articles/stop-a-u-s-coronavirus-outbreak-before-it-starts-11580859525

DEEPFAKES

12 Bethania Palma, "Did President Trump refer to the coronavirus as a 'hoax'?," Snopes. com, 2 March 2020, www.snopes.com/fact-check/trump-coronavirus-rally-remark/

13 https://therecount.com/watch/trump-coronavirus-calendar/2645515793

14 www.youtube.com/watch?v=33QdTOyXz3w

15 Kate O'Keeffe, Michael C. Bender and Chun Han Wong, "Coronavirus casts deep chill over U.S.–China relations," *The Wall Street Journal*, 6 May 2020, available at www.wsj.com/articles/coronavirus-casts-deep-chill-over-u-s-china-relations-11588781420?mod=hp_lead_pos12

16 Alison Rourke and Lily Kuo, "Trump claims China will 'do anything' to stop his re-election as coronavirus row escalates," *The Guardian*, 30 April 2020, available at www.theguardian.com/world/2020/apr/30/trump-claims-china-will-do-anything-to-stop-his-re-election-as-coronavirus-row-escalates

17 Mark Mazzetti, Julian E Barnes, Edward Wong and Adam Goldman, "Trump officials are said to press spies to link virus and Wuhan labs," *The New York Times*, 30 April 2020, available at www.nytimes.com/2020/04/30/us/politics/trump-administration-intelligence-coronavirus-china.html

18 Andrew Romano, "New Yahoo News/YouGov poll shows coronavirus conspiracy theories spreading on the right may hamper vaccine efforts," Yahoo News, 22 May 2020, https://news.yahoo.com/new-yahoo-news-you-gov-poll-shows-coronavirus-conspiracy-theories-spreading-on-the-right-may-hamper-vaccine-efforts-152843610.html?guccounter=1&guce_referrer=aHR0cHM6Ly9jYnNhdXN0aW4uY29tL25ld3M-vbG9jYWwvcG9sbC00NC1vZi1yZXB1YmxpY2Fucy10aGluay1iaWxsLWdhdGVzLXRvLX-VzZS1jb3ZpZC0xOS12YWNjaW5lLXRvLWltcGxhbnQtdHJhY2tpbmctY2hpcA&guce_re-ferrer_sig=AQAAAJJ9Dy7VCfYj_ZOE04qPPXQhCyEQudxJ9E7kNtJ1z6kcuM-oz16M-4PaEA--c3M52QzdBosBDMvTalkUc3fqhEuoBTKBoUAnxLkYehEF1Dubv8pour6uAzWgC-GCYEjPp0BBmwEiWyemMD4Gg891yNugPNDIUjy78yjzsVdHwZTPSp

19 "Covid-19: the psychology of conspiracy theories," *The Guardian* podcast, presented by Ian Sample, www.theguardian.com/science/audio/2020/may/05/covid-19-the-psychology-of-conspiracy-theories

20 Andrew Romano, "New Yahoo News/YouGov poll shows coronavirus conspiracy theories spreading on the right may hamper vaccine efforts," see note 18 above.

21 The anti-vax community is an easy target for foreign hostile states. Indeed, Russia exploits them just as it exploits African Americans.

22 Patrick Clarke, "M.I.A. clears up her stance on vaccinations following Twitter backlash," *NME*, 3 April 2020, available at www.nme.com/news/music/m-i-a-clears-up-stance-vaccinations-following-twitter-backlash-2640812

23 Reuters, "Novak Djokovic's anti-vaccination stance may stop his return to tennis," *The Guardian*, 20 April 2020, available at www.theguardian.com/sport/2020/apr/19/novak-djokovic-coronavirus-covid-19-vaccination-tennis

24 James Temperton, "The 5G coronavirus conspiracy theory just took a really dark turn," *Wired*, 7 May 2020, www.wired.co.uk/article/5g-coronavirus-conspiracy-theory-attacks

25 "XR Belgium posts deepfake of Belgian premier linking Covid-19 with climate crisis," *The Brussels Times*, 14 April 2020, available at www.brusselstimes.com/all-news/belgium-all-news/politics/106320/xr-belgium-posts-deepfake-of-belgian-premier-linking-covid-19-with-climate-crisis/

26 www.facebook.com/watch/ExtinctionRebellionBE/

27 Gideon Rachman, "Jair Bolsanaro's populism is leading Brazil to disaster," *Financial Times*, 25 May 2020, available at www.ft.com/content/c39fadfe-9e60-11ea-b65d-489c67b0d85d

28 https://mrc-ide.github.io/covid19-short-term-forecasts/index.html

29 "Editorial: COVID-19 in Brazil: 'So what?,'" *The Lancet* 395.10235 (9 May 2020), available at www.thelancet.com/journals/lancet/article/PIIS0140-6736(20)31095-3/fulltext

30 Alexander Baunov, "Where is Russia's strongman in the coronavirus crisis?," *Foreign Affairs*, 27 May 2020, available at www.foreignaffairs.com/articles/russian-federation/2020-05-27/where-russias-strongman-coronavirus-crisis

31 www.interpol.int/en/News-and-Events/News/2020/Global-operation-sees-a-rise-in-fake-medical-products-related-to-COVID-19

Chapter 7: Allies, Unite!

1 www.bellingcat.com

2 www.newsguardtech.com/free/

3 www.newsguardtech.com/misinformation-monitor/

4 First Draft News, *First Draft's Essential Guide to Newsgathering and Monitoring on the Social Web* (October 2019); available at https://firstdraftnews.org/wp-content/uploads/2019/10/Newsgathering_and_Monitoring_Digital_AW3.pdf?x14487

5 https://firstdraftnews.org/latest/partnership-on-ai-first-draft-begin-investigating-labels-for-manipulated-media/

6 https://reutersinstitute.politics.ox.ac.uk/about-reuters-institute

7 https://reporterslab.org

8 www.niemanlab.org

9 https://faculty.ai/ourwork/identifying-online-daesh-propaganda-with-ai/

10 https://faculty.ai/ourwork/identifying-online-daesh-propaganda-with-ai/

11 www.darpa.mil/program/media-forensics

12 www.kaggle.com/c/deepfake-detection-challenge/overview/description

13 https://jigsaw.google.com/issues/

14 www.newsprovenanceproject.com/about-npp

15 Hanaa' Tameez, "Here's how *The New York Times* tested blockchain to help you identify faked photos on your timeline," niemanlab.org, 22 January 2020, www.niemanlab.org/2020/01/heres-how-the-new-york-times-tested-blockchain-to-help-you-identify-faked-photos-on-your-timeline/

16 David Smith, "Trump signs executive order to narrow protections for social media platforms," *The Guardian*, 29 May 2020, available at www.theguardian.com/us-news/2020/may/28/donald-trump-social-media-executive-order-twitter

17 Charles Duan and Jeffrey Westling, "Will Trump's executive order harm online speech? It already did," *Lawfare*, 1 June 2020, www.lawfareblog.com/will-trumps-executive-order-harm-online-speech-it-already-did

18 Ron Miller, "IBM unveils blockchain as a service based on open source Hyperledger Fabric technology," techcrunch.com, 20 March 2017, https://techcrunch.com/2017/03/19/ibm-unveils-blockchain-as-a-service-based-on-open-source-hyperledger-fabric-technology/

19 Kathryn Harrison, "Deepfakes and the Deep Trust Alliance," www.womeninidentity.com, 5 November 2019, available at https://womeninidentity.org/2019/11/05/kathryn-harrison-deepfakes-deeptrustalliance

20 "National security concept of Estonia," adopted by the Riigikogu [Estonian parliament], 12 May 2010 (unofficial translation), www.eda.europa.eu/docs/default-source/documents/estonia---national-security-concept-of-estonia-2010.pdf

Image acknowledgements

Page 25 left, samples from the Toronto Faces Database (Susskind, J., Anderson, A., Hinton, G.E.: The Toronto Face Dataset. Technical Report UTML TR 2010-001, University of Toronto, 2010). Image used in Generative Adversarial Nets by Ian J. Goodfellow, Jean Pouget-Abadie, Mehdi Mirza, Bing Xu, David Warde-Farley, Sherjil Ozair, Aaron Courville, Yoshua Bengio. Departement d'informatique et de recherche operationnelle, Universite de Montreal, Montreal, QC H3C 3J7, 2014 © Ian J. Goodfellow 2014. Available at https://papers.nips.cc/paper/5423-generative-adversarial-nets.pdf; page 25 right from A Style-Based Generator Architecture for Generative Adversarial Networks by Tero Karras, Samuli Laine, Timo Aila, 2018/NVIDIA. Available at arxiv.org/abs/1812.04948; page 28 both photos David King Collection © Tate; pages 67, 71, 73 Social Media Advertisements | Permanent Select Committee on Intelligence, available at https://intelligence.house.gov/social-media-content/social-media-advertisements.htm

INTERVIEWEES

Alexander Adam, Data Scientist, Faculty

Andrew Briscoe, Head of EMEA Equity Capital Markets Syndicate, *Bank of America* Merrill Lynch

Areeq Chowdhury, Head of Think Tank, Future Advocacy

Casey Newton, Journalist (Platforms and Democracy), The Verge

Georgio Patrini, Founder and CEO, DeepTrace Labs

Henry Adjer, Head of Threat Intelligence, DeepTrace Labs

Jennifer Mercieca, Associate Professor, Department of Communication, Texas A&M University

Johannes Tammekänd, CEO and Co-Founder, Sentinel

John Gibson, Chief Commercial Officer, Faculty

Kathryn Harrison, CEO and Founder, DeepTrust Alliance

Matthew Ferraro, Counsel, Wilmerhale

Mounir Ibrahim, Vice-President of Strategic Initiatives, TruePic

Samantha Cole, journalist, Motherboard

Sam Gregory, Programme Director, WITNESS

Renee DiResta, Technical Research Manager, Internet Observatory, Stanford University

Victor Riparbelli, Co-Founder and CEO, Synthesia

ACKNOWLEDGEMENTS

While the ideas in this book have been developing for many years, it was commissioned and written at breakneck speed. It was only in January this year that my publisher, Jake Lingwood, asked me to write a book about deepfakes. I am particularly grateful to Jake and the team at Octopus, including Alex Stetter, who worked with me to get this book done in record time. I would also like to acknowledge Martin Redfern, my agent at Northbank Talent, who has patiently been developing this idea with me for over a year. I would also like to thank the many experts who were gracious enough to grant me an interview, often at very short notice, once I had decided to write this book. They were exceedingly generous with their time and thoughts. Many are named in the book, but for those who are not, I am enormously grateful for our many conversations, and for how your input has helped hone my thoughts and observations.

Finally, this book—written mostly during Covid-19 lockdown in spring and early summer 2020, would not have been possible without the help of my "hometeam." This includes Agnes, who has provided incredible and unfaltering support with the care of my baby. The person to whom I owe the most, however, is my partner, James. Not only did he "hold the fort" for months as I was furiously writing, but he has also been essential to the intellectual process. The ideas contained in this book have been honed by their testing on his brilliant mind.

Nina Schick
June 2020